POLITICS OF CLIMATE CHANGE

Crises, Conventions and Cooperation

POLITICS OF CLIMATE CHANGE
Crises, Conventions and Cooperation

Editors

Swaran Singh
Jawaharlal Nehru University, India

Reena Marwah
University of Delhi, India

World Scientific

NEW JERSEY · LONDON · SINGAPORE · BEIJING · SHANGHAI · HONG KONG · TAIPEI · CHENNAI · TOKYO

Published by

World Scientific Publishing Co. Pte. Ltd.
5 Toh Tuck Link, Singapore 596224
USA office: 27 Warren Street, Suite 401-402, Hackensack, NJ 07601
UK office: 57 Shelton Street, Covent Garden, London WC2H 9HE

Library of Congress Control Number: 2022058149

British Library Cataloguing-in-Publication Data
A catalogue record for this book is available from the British Library.

POLITICS OF CLIMATE CHANGE
Crises, Conventions and Cooperation

ISBN 978-981-126-374-3 (hardcover)
ISBN 978-981-126-375-0 (ebook for institutions)
ISBN 978-981-126-376-7 (ebook for individuals)

For any available supplementary material, please visit
https://www.worldscientific.com/worldscibooks/10.1142/13067#t=suppl

Desk Editors: Aanand Jayaraman/Sandhya Venkatesh

Typeset by Stallion Press
Email: enquiries@stallionpress.com

Preface

We need a green planet but the world is on red alert. We are at the verge of the abyss. We must make sure the next step is in the right direction. Leaders everywhere must take action. First by building a global coalition for net zero emissions by 2050 in every country, every region, every city, every company, and every industry.

António Guterres, UN Secretary-General

This volume comprising 13 chapters makes a unique addition to the existing body of literature on Climate Change discourse by thoroughly investigating contemporary issues and initiatives as nations continue to search for novel mechanisms to alleviate, adapt, and mitigate the negative effects of climate change. Especially, the Anthropocene debate has increasingly blurred the distinction between nature and culture, which implies that the natural world cannot be merely seen as an inevitably evolving backdrop to the human history. The Anthropocene perspectives represent a new language of broad-based consensus on climate change, one that has been riddled with divergent interpretations and problematic binaries of utopian and dystopian futures of our beautiful blue planet.

The year 2020 was a watershed event in the history of climate change politics. It marked the end of the second commitment period of the Kyoto Protocol and the beginning of the ambitious Paris Agreement, a nonbinding instrument of INDCs that reflect the key shift and aims to better balance the global politics of climate change. It was the year of the pandemic, which scourged the world and its lingering effects continues to ravage many parts of the world. The disruption caused by this pandemic carries severe implications on a global scale, which range from the return of hunger, poverty, and food insecurity to further exacerbate geopolitics and mistrust. The pandemic has brought before the world the severity and scale of the transboundary challenges in a globally interconnected world. It has especially exposed the weaknesses of our extant global institutions and governance structures and processes meant to tackle the complex and imminent threats from climate change.

In this rapidly evolving backdrop this volume titled, *Politics of Climate Change: Crises, Conventions and Cooperation* seeks to bring to the reader, an array of evolving discourse that have catapulted Climate Change into the very center stage of discussions ranging from development to diversity, from economy to ecology and from technologies to transformation. And, as States prepare for the future of global climate change negotiations and initiatives post the Glasgow COP26 event of November 2021, there has been a significant visible shift in the politics of climate change at various levels. The negotiations that took place in the shadows of the pandemic have managed to challenge the political lethargy and non-committal attitudes of states on the climate change question.

Unlike in the past, climate change has now come to be a key issue on the political high tables of all hues and colors. This issue has spilled outside COP negotiating spaces and into the public sphere. The politics of climate change, which was once limited to the UNFCCC and COP events, has emerged to be a far wider global concern. Whether it is the school strikes led by children or the indigenous struggles of the marginalized populations, the politics of climate change today is far more diverse, representative, and hyperactive. At the same time, we can witness the shifts in the state's understanding of the problem as well, which are actively inquiring about its security and geopolitical dimensions. The boundaries between traditional and non-traditional threats to security are getting blurred as climate change and its myriad impacts wreak havoc on ecosystem resilience, state's welfare capacity and everyday lives of people. The latest Intergovernmental Panel on Climate Change (IPCC) Report of 2021 (which is the sixth assessment report and has come after eight years of the previous one) reports that irreversible effects on the ecosystem and of added vulnerabilities of mitigation and adoption is expected to kickstart serious engagements at various levels.

As editors of this volume, we do hope that governments, institutions, and individuals will work together to reduce carbon emissions which threaten life on land, in water and even in space. We believe, this is every individual's battle to save humanity from extinction and the time to take action is now!

Prof. Swaran Singh
Prof. Reena Marwah

About the Editors

Swaran Singh is Chairman and Professor, Centre for International Politics, Organisation and Disarmament (CIPOD), School of International Studies, Jawaharlal Nehru University (New Delhi), President of Association of Asia Scholars (New Delhi), Member, Governing Body, Society of Indian Ocean Studies (New Delhi). Singh has been formerly visiting professor/scholar at Australian National University (Canberra), Science Po (Bordeaux, France) University of Peace (Costa Rica), Peking, Fudan, and Xiamen Universities, and Shanghai Institute of International Studies and Center for Asian Studies (Hong Kong University) in China, Asian Center (University of the Philippines), and Chuo, Hiroshima, and Kyoto Universities (in Japan), as also Guest Faculty at Stockholm International Peace Research Institute (Sweden). He was Academic Consultant (2003–2007) at Center de Sciences Humaines (New Delhi), Research Fellow, Institute for Defence Studies and Analysis (New Delhi).

Singh has published in *Journal of International Affairs* (Columbia University), *Security Challenges* (Australian National University), *Journal of Indian Ocean Region* (Perth, Australia), *Issues & Studies* (Taiwan National University), *African Security* (Institute of Security Studies), *BISS Journal* (Dhaka), and several Chinese and Indian journals. Singh co-edited *Multilateralism in the Indo Pacific — Conceptual and Operational Challenges* (Routledge, 2022), *Revisiting Gandhi: Legacies for Global Peace and National Integration* (World Scientific, Singapore, 2021), *Corridors of Engagement* (2020), *Colonial Legacies and Contemporary Studies of China and Chineseness: Unlearning Binaries* (2020), *BCIM Economic Corridor: Chinese and Indian Perspectives* (2017), *Transforming South Asia: Imperatives for Action* (2013); *India and the GCC Countries, Iran and Iraq: Emerging Security Perspectives* (2013), *On China by India: From Civilization to State* (2012), *Emerging China: Prospects for Partnership in Asia* (2012), *Asia's Multilateralism* (in Chinese 2012); Edited *China–Pakistan Strategic Cooperation: Indian Perspectives* (2007); Co-authored *Regionalism in South Asian Diplomacy* (2007) and authored *Nuclear Command & Control in Southern Asia: China, India, Pakistan* (2010), *China–India Economic Engagement: Building Mutual Confidence* (2005), *China–South Asia: Issues, Equations, Policies* (2003).

Singh has supervised 32 PhDs and 50+ MPhil degrees at JNU and sits on Selection Committees for faculty recruitment and on the Editorial Board of various reputed journals. He regularly writes for Indian and foreign media, lectures at various prestigious institutions in India and abroad, and regularly appears on radio and television discussions. Twitter: @ SwaranSinghJNU.

Reena Marwah (MPhil, Delhi University; PhD, India, International Business) is Professor at Jesus and Mary College, Delhi University.

She was an ICSSR Senior Fellow, MHRD, Govt. of India, affiliated with the Centre for the Study of Developing Societies, New Delhi from June 2017 to May 2019, during which her study was on Reimagining India–Thailand Relations. She has also been on deputation as Senior Academic Consultant, ICSSR, Ministry of Human Resource Development, Govt. of India for three years (2012–2015) and continued, on behalf of ICSSR to coordinate/ lead the India–Europe Research Platform (EqUIP), comprising 10 research councils of Europe till July 2017. She is the recipient of several prestigious fellowships, including the McNamara fellowship of the World Bank, 1999–2000, and the Asia fellowship of the Asian Scholarship Foundation 2002–2003, during which she undertook research in Thailand and Nepal. She is also a Senior Fellow of the Institute of National Security Studies Sri Lanka (INSSSL). She has been a Consultant for the World Bank and UN Women. She is the founding editor of *Millennial Asia*, a triannual journal on Asian Studies of the Association of Asia Scholars, published by Sage Publishers.

During her teaching and research experience, she has worked closely with several think tanks, international donors, embassies, ministries of the Government of India, and research councils in Asia. Among her research interests are international relations issues of China, Vietnam, Philippines, Thailand and India, and development issues of South and South East Asia.

In addition to several chapters and articles published in books/journals, she is author/co-author/co-editor of 16 books and monographs, including *Contemporary India: Economy, Society and Polity* (Pinnacle 2009, 2011), co-edited volumes including *Economic and Environmental Sustainability of the Asian Region* (Routledge 2010), *Emerging China: Prospects for Partnership in Asia* (Routledge 2011), *On China by India: From a Civilization to a Nation State* (Cambria Press, USA), *Transforming South Asia: Imperatives for Action* (Knowledge World, 2014), *The Global Rise of Asian Transformation* (Palgrave Macmillan, 2014).

Her latest co-edited books are: *China Studies in South and Southeast Asia: Pro-China, Objectivism, and Balance* (World Scientific, 2018), *Revisiting Gandhi: Legacies for Global Peace and National Integration* (World Scientific, Singapore, 2021) and *Multilateralism in the Indo Pacific — Conceptual and Operational Challenges* (Routledge, 2022).

Her most recent authored books are *Re-imagining India–Thailand Relations: A Multilateral And Bilateral Perspective* (World Scientific, March 2020), *China's Economic Footprint in South and Southeast Asia: A Futuristic Perspective* (World Scientific, 2021) and *India–Vietnam Relations: Development Dynamics and Strategic Alignment* (Springer Nature, 2022).

About the Contributors

Aditi Basu is an independent researcher who has completed her Masters' in Political Science from Jamshedpur Women's College, Jharkhand, India. Her research interests include Indian Foreign Policy, International Relations and Comparative Politics.

Anand Sreekumar pursued an MPhil in diplomacy and disarmament at the Centre for International Politics, Organization and Disarmament in the School of International Studies, Jawaharlal Nehru University. He holds a Masters' degree in Development Studies from the Indian Institute of Technology, Madras. His research interests lie at the intersection of IR theories, climate change, and nuclear politics as well as South Asian political thought.

Artyom A. Garin is Fellow at the Center for Southeast Asia, Australia and Oceania at the Institute of Oriental Studies of the RAS. He is interested in multilateralism in the Indo-Pacific, as well as in Australia–China relations. His research interests also include defense and aid policies of Australia, as well as politics and history of the Pacific Island Countries.

Chaitra C. (PhD) is Assistant Professor and Head of the Department of Political Science, Govt. First Grade College, C.S. Pura, Karnataka. She has been an e-content developer on various topics of Political Science for the Govt. of Karnataka. Her areas of interest are climate change and governance, political theory and thought.

Claudia Astarita is lecturer at Sciences and research associate at the Lyon's Institute of East Asian Studies in Lyon, France. She obtained her PhD in Asian Studies from Hong Kong University and her main research interests include China's political and economic development, Chinese and Indian Foreign policies, East Asian regionalism and regional economic integration, Asian Civil Society, and the role of media and memory in reshaping historical narratives in Asia.

Julius Hulshof, having completed his undergraduate studies in anthropology and political science at University College Utrecht, recently finished the dual Master's program in International Relations at Sciences Po and PKU.

Kakoli Sengupta is Associate Professor and Former Head of Department, Department of International Relations, Jadavpur University, Kolkata, West Bengal, India. Her areas of specialization are Terrorism, Counter Terrorism, Peace & Conflict, Gender, European Security and Irish politics. She appears regularly on television programs on topics of current affairs and international relations.

Kunwar Alkendra Pratap Singh (PhD) is Assistant Professor in Department of Physics, Banaras Hindu University. He did his postdoctoral research from Kyoto University, Japan and Indian Institute of Astrophysics, Bangalore. His research is published in peer-reviewed journals such as *Astrophysical Journal Letters* (ApJL), *Astrophysical Journal* (ApJ), etc. His current work focuses on Astronomy and Outer Space Security.

Md Abid Hasan is a development professional and independent researcher with interest and expertise in entrepreneurship, WASH, sustainability, and climate change. Currently, he is involved with Vision Green Organization as a consultant. Abid has pursued both of his Bachelor and Masters' degrees in Development Studies from the University of Dhaka.

Namit Mahajan is an undergraduate student of the Department of Economics, Ramjas College, University of Delhi.

Oliver Tirtho Sarkar is a Lecturer in the Department of Environment and Development Studies at United International University. His theoretical and empirical experience spans academia, research, national and international development sectors. He pursued his undergraduate and postgraduate degrees from the University of Dhaka and majored in Development Studies.

Prathit Singh is an undergraduate student of the Department of Political Science, Ramjas College, University of Delhi.

Rabby Us Suny is currently working as an "Annotation Specialist" in Augmedix Bangladesh Limited. He has obtained Bachelors' of Social Science and Masters' of Social Science degrees from the Department of Peace and Conflict Studies, University of Dhaka and began his career in the research sector. He is a social-research enthusiast possessing noteworthy skills in qualitative research.

Ritvick Khanna is an undergraduate student of the Department of English, Ramjas College, University of Delhi.

Saurabh Thakur (PhD) is a Consultant at the Ministry of External Affairs, India. His research sits at the intersection of climate governance, international politics and sustainability, looking specifically at the climate security issues in the context of South Asia. Previously, he was associated with National Maritime Foundation, New Delhi, Coalition for Disaster Resilient Infrastructure (CDRI) and the Regional Centre for Strategic Studies, Colombo, Sri Lanka. He holds a PhD degree in International Relations from the Centre for International Politics, Organization and Disarmament (CIPOD), JNU, India.

Sheeraz Ahmad Alaie (PhD) is presently working as a DST-STIP Postdoctoral Fellow hosted jointly by DST, New Delhi and DST-CPR Panjab University Chandigarh. He completed his PhD in Science, Technology and Innovation Policy from Central University of Gujarat in 2019. His areas of research are Innovation and Innovation System, Science Policy, Agricultural Innovation System, Public Private Partnerships and Sustainable Development.

Sonia Roy is Assistant Professor of Political Science, West Bengal Education Service, presently posted at Taki Government College (North 24 Parganas, West Bengal).

Swasti Rao (PhD) is Associate Fellow at the Europe and Eurasia Center, Manohar Parrikar Institute for Defence Studies and Analyses. She has completed her PhD in Advanced International Politics from Tsukuba University, Japan. She is a recipient of Japanese Education Ministry Fellowship (MEXT) in 2011 and also a recipient of the "Okita Memorial Scholarship" in 2010 from ICCR.

Acknowledgments

This volume titled, *Politics of Climate Change: Crises, Conventions and Cooperation* is an outcome of a three-day International Conference held in a hybrid mode in India Habitat Centre and online in November 2021. The conference inaugural and keynote addresses were delivered by Prof. Ilan Kelman Professor of Disasters and Health, University College London, United Kingdom and Prof. Sanjay Chaturvedi, Dean, Faculty of Social Sciences, South Asian University, New Delhi, respectively. The editors are deeply grateful to both experts for elucidating the context and contemporary climate change nuances. Their addresses helped to steer the deliberations which followed on two subsequent days.

The conference brought together authors from varied disciplines to focus on the range of issues encompassed within the theme of climate change. Authors submitted their complete papers weeks prior to the conference to enable discussants to review and comment on the selected research papers. The authors were then required to substantively revise their papers as chapters based on the comments received from the discussants during the conference as well as comments received from Editors. This volume acknowledges the perseverance of several authors whose papers were revised few times and all of them have contributed to enriching the book.

Editors also take this opportunity to thank each of the conference session chairpersons and discussants, whose valuable inputs helped to enrich the contributions of the authors. We are particularly grateful to Prof. Lakhwinder Singh and Prof. Sukhpal Singh for chairing various sessions. Our thanks are also due to a large number of scholars who participated in this three-day conference and engaged the presenters with pointed questions. Conference participants are also acknowledged for their candid sharing of views.

We are also indebted to Prof. Sachin Chaturvedi Director General at Research and Information System for Developing Countries (RIS) for an insightful Valedictory Address as well as to Dr. E. Sridharan, Academic Director and Chief Executive of the University of Pennsylvania Institute for the Advanced Study of India (UPIASI) for ably chairing the Valedictory session.

This volume comprising 13 chapters would not have been possible without the kind cooperation of the production and editorial team of World Scientific Publishers, Singapore. Each one deserves our sincere thanks and appreciation. All research interns of our Association of Asia Scholars, led by Dr. Silky Kaur, Dr. Saurabh Kumar and Dr. Chaitra C. were continuously engaged in ensuring the success of the conference and deserve our appreciation. Finally, we are grateful to our families for being our constant strength in enabling us to complete this seminal work.

Prof. Swaran Singh
Prof. Reena Marwah

Contents

Preface   v

About the Editors   ix

About the Contributors   xiii

Acknowledgments   xvii

Abbreviations   xxi

Chapter 1   Politics of Climate Change: Accords and Discord   1
             Swaran Singh and Reena Marwah

Section I   Issues   **23**

Chapter 2   All Ships Are Not Raised: The Politics of Climate
             Disasters in the Anthropocene   25
             Saurabh Thakur

Chapter 3   Climatic Politics of Non-State Actors in the
             Post-Pandemic Era: Insights from Eco-Cinema
             of the Global South   45
             Anand Sreekumar

Chapter 4   Climate Change as a New Area of Sino-Quad
             Competition: Pacific Islands Perspectives   67
             Artyom A. Garin

Chapter 5   Political Economy of River Ecocide in Bangladesh:
              A Study in the Context of Dhaleshwari River   83
              Rabby Us Suny, Oliver Tirtho Sarkar,
              and Md Abid Hasan

Chapter 6   Challenges of Space Debris and Space Drag:
              Building an International Climate Change Regime   105
              Swasti Rao and Kunwar Alkendra Pratap Singh

Section II   Institutions and Initiatives   **129**

Chapter 7   Climate Action by the European Union: Making
              the European Green Deal a Reality   131
              Kakoli Sengupta

Chapter 8   International Solar Alliance: Testing a New
              Framework to Approach Energy Shortage   149
              Claudia Astarita and Julius Hulshof

Chapter 9   Multilateralism Efforts in Asia: What's the
              Way Forward?   169
              Prathit Singh, Namit Mahajan, and Ritvick Khanna

**Section III   Climate Change Narratives with a Focus
              on India**   **185**

Chapter 10   International Climate Governance: Indian
              Perspectives   187
              Chaitra C.

Chapter 11   Policies as Instruments in Promoting Sustainable
              Development: Limiting the Climate Change Issues
              in India   201
              Sheeraz Ahmad Alaie

Chapter 12   The Impact of Fridays for Future Movement
              in Indian Politics   217
              Sonia Roy

Chapter 13   Emerging Indian Partnerships in Climate Change
              with Special Reference to COVID-19 Era   235
              Aditi Basu

Index   253

Abbreviations

AR4	Fourth Assessment Report of the Intergovernmental Panel on Climate Change
AR5	Fifth Assessment Report of the Intergovernmental Panel on Climate Change
CO_2	Carbon dioxide, a greenhouse gas
CBDR	Common but Differentiated Responsibilities
CDM	Clean Development Mechanism of the Kyoto Protocol
COP21	21st meeting of the Conference of the Parties
EU	European Union
GDP	Gross Domestic Product
IPCC	Intergovernmental Panel on Climate Change
PICs	Pacific Island Countries
REDD+	Reducing Emissions from Deforestation and Forest Degradation in Developing Countries
UNEP	United Nations Environment Programme
UNFCCC	United Nations Framework Convention on Climate Change
UNCHE	United Nations Conference on the Human Environment
WRI	World Resources Institute

https://doi.org/10.1142/9789811263750_0001

Chapter 1

Politics of Climate Change: Accords and Discord

Swaran Singh and Reena Marwah*

Abstract

This chapter provides insights into varying perspectives on climate change which underpin the politics circumventing this subject. Several reports and experts have discussed the pitfalls of politicking around a subject as overpowering and overwhelming as climate change which has been the outcome of humanity's unrelenting assault on the very resources which support life on this beautiful blue planet, Earth. Evidently, countries have been evading responsibilities as the pollutants that cause climate change mix across national borders. This blue planet has been warming at an unprecedented rate; especially so in the last 50 years and cooperative actions and mechanisms with the political will to implement these must be the way forward. However, whether governments will cooperate by investing in institutions and technologies and provide required financial support needed to reduce the carbon footprint of human actions depends on their chosen priorities. Countries clearly vary in population, affluence, technology as also in terms of their relative vulnerability to climate impacts — factors that, among others, affect how much they are willing

---

* Swaran Singh teaches diplomacy and disarmament at Jawaharlal Nehru University and is president of Association of Asia Scholars (asiascholars.in). Reena Marwah teaches international business in Delhi University.

to pay to address global climate change. Hence, in the first section this chapter interrogates the capacity of governments to pledge their commitment to climate action. The second section comprises the structure of the volume as it aims to deepen the debate and understanding of contemporary climate change narratives.

Keywords: Climate Change, Mitigation, Adaptation, Investment, Politics

Introduction

The environment has been a subject of attention for several decades; however, the issue has been politicized on the international agenda only since 1980s. It was in the mid-1980s when the issue received attention from scientists, media, the environmentalists, and soon by politicians. The Toronto International Climate conference of 1988 was the first global initiative to produce a non-binding agreement that emissions be reduced by countries. The urgency to understand and address exigencies of climate change was hastened by a series of extreme weather phenomena in the late 1980s, especially in the United States (US). These weather events were interpreted by environmentalists and others as potential harbingers of climate change and massive media coverage increased political pressures to initiate concrete action at national level. Despite this, several other issues such as the burning of fossil fuels, consumption of energy resources among others have since received a tepid response both from the developed and developing countries. COP26 held in November 2021 underlined the imperative for nations to translate their pledges into rapid action and reduce dependence on fossil fuels. However, the broken promise of $100 billion a year by the developed countries for climate action since 2009 remains unfulfilled thereby undermining trust in developed countries' commitment to climate multilateralism.

This chapter delves into the politics of climate change and helps unfold this dilemma where nations have largely failed to implement their pious sentiments repeated *ad nauseam* year after year at multiple forums. The manner in which climate change has been politicized by nations forms the essence of this volume. The authors also seek to elucidate various milestones and main conflicts that have determined the pace and impasse in international negotiations; even as well-organized interest groups and science-based institutes and international organizations have had their own share of contradicting agendas and open conflicts. The authors here attempt to also identify the causative factors which explain the difficulties and conflicts, combined with the process for achieving an effective and binding agreement at the global level. More recently, the COVID-19 pandemic and many zoonotic and other infectious diseases since early 2000s have also had origins in environmental degradation and climate change (O'Callaghan-Gordo and Antó 2020). It is evident then that the response to the pandemic (which has overwhelmed nations since early 2020) must understand its linkages and also ramifications of climate change on the overall health of planet Earth. Public policies, must be geared toward preservation of natural ecosystems as well as the many stakeholders of the system including economic actors, the financial sector as well as producers and consumers.

In the wake of coronavirus pandemic, the World Health Organization has identified that diseases which can be transmitted to humans from animals emanate due to a decrease in immunity of human beings as well as deforestation and the consumption of meat of different animals (WHO 2012). This pandemic has reinforced interconnectedness of our blue planet that impels shared efforts with nations with added leverages taking lead in strategies of adaptation and mitigation in making finances, technologies, and skillsets to those in need. The present patterns of consumption as well as the structures of production are becoming increasingly environmentally unsustainable. The role of sustainable development cannot be hence overemphasized. Especially in face of the coronavirus pandemic, experts have highlighted how unrestrained trade and consumption of wildlife must be regulated to lower the risks posed by zoonotic diseases (Crossley 2020). Its linkages with climate change have also been revealed by recent studies.

Another important revelation of this increasing focus on studying climate change from multiple perspectives has been one of intense interlinkages that exist among issues and initiatives for development, consumption, production with regard to climate change and environmental degradation which makes it a multidisciplinary issue requiring commitment, cooperation, and concerted actionable goals by the global community as a whole (Kedia, Pandey and Sinha 2020). However, climate change discourse has remained vulnerable to the North–South divide which is clearly visible both in political will and commitment of resources for concomitant actions. The most visible binary highlighted by experts reveals mainstream narratives being captured by the developed North as an instrument of control of the Global South and to possibly thwart the developmental impulses of the developing nations. Experts have argued for an imperative for greater understanding of the environment as the globalized integrated whole even when politics continues to sustain profound asymmetries in human development. Nation states, therefore, must accept responsibility for holistic environmentally friendly policies which, while protecting the environment do not impede human progress (Chaturvedi and Doyle 2015).

IPCC Report 2022: Cold Truths for a Warming Planet

The much-awaited Intergovernmental Panel on Climate Change (IPCC) sixth report on climate science, an outcome of Working Group III on climate change mitigation, was published on April 4, 2022. The report explains that places with "poverty, governance challenges and limited access to basic services and resources, violent conflict and high levels of climate-sensitive livelihoods" are more vulnerable to climate change impacts. This report also documents the actions humanity can take to reduce greenhouse gas (GHG) emissions and remove carbon dioxide (CO_2) from the atmosphere. Moreover, it cautions nations that global action is far off track as total GHG rose consistently over the previous decade (IPCC 2022).

This report and the dangers that it documents and extrapolates is a most apt sequel to the two weeks of intense discussions, debates, and protests at Glasgow COP26 that concluded on November 14, 2021. This is expected to compel nations to commit to speed up their efforts at emission cuts. However, this does not show promise of nations moving at a pace that can avoid the worst irreversible impacts of climate damage; some of these are already visible around the world. Although nations have kept alive their commitment to the template of 1.5°, it is evident that this can only survive if promises are kept and pledges translated into rapid action, catapulted by the availability of green finance. The broken promise of $100 billion a year by the developed countries for climate action in 2009 has put at stake, "trust in multilateralism" that hits at the core of climate change optimism of few nations. This issue has been repeatedly flagged by India and seconded by several developing countries as also various organizations even inside developed nations. But politics has continued to intervene in limiting choices made by the nation states.

According to Ilan Kelman, the IPCC report completely avoids the phrase "natural disaster" which reflects limitations even in semantics of these debates. In his view, there is sufficient documented evidence that "disasters are caused by sources of vulnerability — such as unequal and inequitable access to essential services like healthcare or poorly designed or built infrastructure like power plants — rather than by the climate or other environmental influences". Second, he also underlines the fact that disasters are not caused by climate change alone; in fact, vulnerability of victims of such disasters exists before any such humanitarian crisis are revealed to trigger initiatives (Kelman 2022). Again, it is the developing and least developed nations that remain increasingly vulnerable to all the negative impacts of climate change. For instance, in the wake of climate change, extreme weather events like heatwaves, floods, and drought have become both far more intense as also more frequent causing disruptions in production, accessibility, and consumption patterns. According to Christina Lu, "The vast majority of the people facing hunger — roughly 80% — reside in regions that are susceptible to these climate extremes" (Lu 2022).

The President and CEO of the World Resources Institute, Ani Dasgupta, likewise elucidates the above argument when saying: "The newest IPCC report confirms what we feared: our window to limit warming to 1.5°C is extremely narrow, but the world's leading scientists have shown us how to keep that goal alive. It's time world leaders wake up and realize that business as usual is a recipe for total disaster". He adds that "This report is an urgent call to action to peak GHG emissions before 2025 and nearly halve GHG emissions by 2030 to set the world on a path to reach net-zero by around mid-century. The sixth IPCC Assessment Report finds that emissions have continued to rise and that current climate plans and policies are woefully inadequate to change this trajectory. But, at the same time, the IPCC report also points to some signs of progress that we can build upon. At least 24 countries have peaked emissions and sustained GHG reductions for over a decade. The rapid uptake of wind and solar has vastly outpaced what scientists previously thought possible, and last year renewable energy accounted for 90% of new power supplies globally". The urgency to hold temperatures and the way forward has also been reiterated repeatedly.

To quote Dasgupta:

We have cost-effective options available today to significantly reduce emissions in the near-term while meeting our development goals.

We must decarbonize electricity generation, industry, transport and buildings while halting and reversing deforestation. We must invest in innovation while closing the climate finance gap. We must improve agricultural practices while reducing food loss and waste and shifting to more sustainable diets. And we must stop investing in new fossil fuel infrastructure while scaling-up carbon removal. (WRI 2022)

India and Climate Change Narratives

India being the world's third largest polluter after China and United States and its carbon emissions rising faster than these two nations have made it both the focus as also home for rigorous climate change analysis and initiatives (Singh 2019). India's carbon emissions may have slowed for the period between 2016 and 2019 compared to 2011–2015, yet set against the world average, it remains higher. More recently, the pandemic did reduce these emissions for all major manufacturing nations yet India's emissions continue to remain higher that world averages (Koshy 2021). Moreover, additional challenges to controlling India's carbon emissions also flow from projections of India being projected to emerge as world's third largest economy as soon as 2030 (Sharma 2022). It is in this context that this volume makes an attempt to analyze various components and contours of climate change discourse from Indian perspectives. It is of course also important here to briefly highlight India's engagement with these narratives from the official perspective which informs much of this volume's analysis but are not directly included in this volume.

To begin with, India remains not just the third largest country for carbon emissions but also its victim. The Global Climate Change Index 2022, for instance, lists India as 14th most vulnerable country. The Global Food Policy Report 2022 projects climate change being responsible for nearly 74 million people going hungry by 2030 with temperature increase resulting in agricultural productivity decreasing by 1.8–6.6% by mid-century (2041–2060) and by 7.2–23.6% by end-century (2061–2080) (IFPRI 2022:130) The aforementioned IPCC Assessment report talks of India having seen 35% growth in urbanization during 2005 and 2020 creating "unique challenges" for India's climate change policy where its "window of opportunity to deal with the crisis is closing" (Deutsche Welle 2022). All this clearly makes India's national leaders responsible not just toward its domestic constituencies but also toward the international community as a whole. India's enormous dependence on import of gas and oil as also inordinate use of coal makes the challenge of cutting on carbon emissions daunting; posing difficult policy choices. Being the world's largest democracy and emerging economy, India's "aspirations as a responsible power require not just that India is part of the solution but is seen to be part of the solution" (Dubash 2019). This marks the backdrop of official India's commitment to being part of climate change strategies.

India's initiatives like International Solar Alliance (ISA), Coalition for Disaster Resilient Infrastructure (CDRI), and several other benchmarks have brought India to the forefront of various mitigation and adaption strategies. Nevertheless, there are voices in India that continue to blame colonial times of extraction policies of major European industrialized nations and their callous industrialization during the last 200 years especially so when they expect all nations, rich and poor, to together achieve zero carbon emission by the middle of the 21st century. No doubt India has moved along with the larger shift from Common but Differentiated Responsibility (CBDR) to Independently Determined National Cut-offs (INDCs) as the basis of national mitigation and adaptation strategies. Yes, as recent as in the G20 Summit of 2021, India had drawn attention to its population size saying "India's per capita greenhouse gas (GHG) emission is around 1.96 tCO2e (ton carbon dioxide equivalent) which is less than one-third of the world per capita GHG emission (6.55 tCO2e)" (Mohan 2021). India has also been vocal about industrialized countries having failed to keep their promises about climate finance and technology transfers to support latecomers into industrialization. But increasingly official India has come to be recognized for initiating substantial efforts and be seen as an important player in global climate change discourses.

In the run up to 2021 Glasgow COP26, for instance, developing countries like India were being pressurized to enhance their climate ambition and define their pathways to achieving their independently stated goals. Among these India, being the third largest polluter was expected to take the lead in showcasing its commitment to further intensify its emission cuts. While India has been appreciated for its initiatives, especially for International Solar Alliance it was also expected to further cut short its expressed timelines. This is where India's Prime Minister Narendra Modi presented to the COP26, India's proposal of five nectar elements, "Panchamrit" (Press Information Bureau 2021). These five elements aim for (i) non-fossil energy capacity to increase to 500 GW by 2030, (ii) meeting 50% of energy requirements from renewable energy by 2030, (iii) reducing the total projected carbon emissions by one billion tons till 2030, (iv) reducing the carbon intensity of its economy by less than 45% by 2030, and (v) achieving the target of New Zero by the year 2070.

This "Panchamrit" speech as part of India's Azadi ka Amrit Mahotsav was presented as grounded in India's ancient traditions and its ongoing assurance about its seriousness in taking lead in combating the negative impacts of climate change. Prime Minister Narendra Modi, of course, did not forget to once again call out the rich nations on their slack action. He reiterated the imperative for climate finance to the extent of $1 trillion, hence juxtapositioning progress in climate mitigation with climate finance. Indian Environment Minister Bhupender Yadav, reiterated this linkage with climate finance while outlining India's commitment on phasing out coal-fired power plants. He underlined that developing countries were "entitled to the responsible use of fossil fuels". He also pointed out India's own contribution to climate action, epitomized through the creation of the International Solar Alliance (ISA), the Coalition for Disaster Resilient Infrastructures (CDRI) as well as the One Sun, One World, One Grid (OSOWOG) initiatives. He criticized the lack of consensus on key issues and blamed "unsustainable lifestyles and wasteful consumption patterns" in rich countries for causing global warming.

While the blame game and finger pointing is part of politics that may continue to the next and subsequent climate conferences, there is an increasing realization about the urgency with which each nation must act for its own survival and sustenance. India, for instance, has made progress in mitigation and adaptation for overcoming negative impacts of climate change; yet there remain areas where India is found lacking. For instance, India has yet not signed onto the pledge for deforestation and tree cover and it must not shy away from accepting that rapid loss of forest cover adversely impacts livelihoods and environment. It may also reconsider joining the Global Methane Pledge for cutting down the emission of this lethal greenhouse gases. After all, as documented by a recent report of the CEEW, the frequency and intensity of extreme climate events in India have increased by almost 200% since 2005" (Mohanty 2020). In May 2022, India and 195 other countries have pledged to accelerate the restoration of a billion hectares of degraded land across the world by 2030. Many more such specific initiatives are required by the global community. It is a truism that climate is a great equalizer and spares no one. The trouble is that we all think we have time!

Structure of the Volume

This volume is a result of multiple revisions and recalibration starting from abstracts being invited for a conference followed by two workshops. Papers presented at the final conference were revised and edited and final version comprises a total of 13 chapters including this introduction chapter while the remaining 12 chapters are covered within three sections titled, Issues, Institutional Initiatives, and Climate Change narratives with a focus on India.

Part 1: Issues

There is no escaping the fact that climate change affects everything from geopolitics to economies to migration. It shapes cities, life expectancies, and food security. With rising temperatures, and climate crisis indicators breaking records in 2021, there are ramifications which must be outlined and ascertained. Hence, the first section of the volume with five chapters provides the reader with an understanding of the perils faced by humanity.

Saurabh Thakur, in Chapter 2 titled *All Ships Are Not Raised: The Politics of Climate Disasters in the Anthropocene*, elucidates how we today live in an age of reflexive modernity, where societies across the world have become susceptible to unknown and unforeseen risks. According to him, this fear of the unknown transforms the complex notions of basic human needs and well-being into a simple hedging against the future risks. He showcases how the contemporary climate change crises are fundamentally different from our classical understanding of disasters, which refer to events whose impacts are restricted in time and space. The Anthropocene, he says, blurs the distinction between the Nature and Culture, which implies that the natural world cannot be merely seen as a gradually evolving backdrop to the human history. The Anthropocene framework represents a new language of consensus building on climate change, one that is riddled with divergent interpretations and problematic binaries of utopian and dystopian futures. Writing about the Paris agreement, he highlights two contradictory developments which will shape the future of climate action — first, the role of scientific expertise is undergoing a massive change in the politics of climate change. This relationship between politics and science, which remained "inconsistent and apocryphal" in the past, has witnessed a shift wherein climate science has transformed from advocacy to a "solution-and future-oriented" regulatory science. The scientific expertise is, in fact, increasingly being viewed as a cogent alternative to political legitimacy. The second development is the emergence of climate realism and waning of the political currency of principles.

This pragmatism of politics, for him, explains the shift from the long-relied binding principle of Common-but-differentiated Responsibilities (CBDR) which had previously shaped the climate regime to principle of Intended Nationally Determined Contribution (INDC) that is fluid and non-binding. This shift for him marks the tragedy of commons on a planetary scale and will be hard to avoid unless the ideological roots of present-day truths are explored and rendered visible. The language of boundaries and limits which dominate the Anthropocene discourse tends to collide with the narratives of development and global distributive justice, which remain dominant in the Global South. This chapter argues that an ahistorical framing of the Anthropocene presents an analytical challenge to the disciplines such as International Politics, where the collision of world history with planetary history will be most visible. Thakur argues that the politics of climate change is increasingly driven by dystopian/utopian futures and it tends to obfuscate the questions of moral culpability, accountability, asymmetry of state power and justice, which remain the dominant concerns in the present world.

The next chapter by **Anand Sreekumar**, titled, *Climatic Politics of Non-state Actors in the Post-pandemic Era: Insights from Eco-cinema of the Global South* highlights how the recent climate change discourse, in the midst of Paris Agreement and the coronavirus pandemic, have been shaped by myriad manifestations of the politics of Eco-cinema. While the sites of politics have rightly percolated from the high politics of states to the low politics of non-state actors, the literature on the latter is overwhelmingly dominated by civil society, grassroots movements, etc. He stresses that it is unfortunate that the politics of non-state actors like filmmakers has been largely overlooked in climate change discourse. In what he calls the post-literate age where visual media has emerged as the supreme source of information, eco-cinema has come to be a key site of representation of climate change. Backed by a theoretical framework of visibility strategy, he deploys the methodology of a comparative visual content and discourse analysis to juxtapose eco-cinema from the Global North (*Corals, Our Planet*, etc.) and Global South (*Kadvi Hawa, Nelson,*

The Weeping Apple Tree, etc.) covering a wide spectrum of visual media, from documentaries to movies, representing diverse experiences of climate change across the world. Based on this comprehensive analysis, he argues that the politics of filmmakers in the Global North is shaped by "pre-trauma" anticipating a future dystopia, fetishization of charismatic megafauna, extreme climatic events, etc., expressed through an excessively cinematic language of shock value. However, their counterparts from the Global South, while sharing important convergences like trauma/loss, diverge on the question of the temporality of trauma as situated in the mundane everyday life; in the backdrop of socio-political injustices, echoing a post-disaster realism. This signifies more nuanced stories of suffering, adaptation, and resilience mirroring the daily realities of the most marginalized people who are the most affected. In conclusion, he argues that the pandemic, exacerbating climate-change insecurities across the world, has provided an opportunity to reframe eco-cinematic representations of climate change by non-state actors in terms of equity and social justice as opposed to depoliticized spectacles.

Taking the discussion to a more contemporary arena, **Artyom A. Garin's** chapter titled, *Climate Change as a New Area of Sino-Quad Competition: Pacific Islands Perspectives*, highlights how the combination of natural resources, unique geographical position, extensive EEZs, and the challenges facing the Pacific Island Countries (PIC) — including economic and natural shocks — have made climate change an urgent issue for the region, which have a negative impact on PICs' critical infrastructure and food security. However, there is another dimension of climate change — geo-strategic. The intensification of the Sino-U.S. (incl. Quad) competition for leadership in the Indo-Pacific forces them to engage in new areas of cooperation with small states in order to gain advantages in the region. PICs are no exception. China, Australia, New Zealand, the United States, Japan, or Taiwan have been using Official Development Assistance (ODA) for decades to involve or keep PICs in their orbit. Australia, as the traditional leader of the South Pacific, has already secured its economic, defense, and vaccine presence in PICs. But the climate agenda, despite PICs' attempts to attract the attention of "senior" partners, could become a new platform for geo-strategic competition.

An important role in this climate change-driven geo-strategic competition has been played by the assertive commitment by the U.S. President Joseph Biden's administration to the global climate action agenda which will affect the foreign policy vector of Australia in the same way that Donald Trump's foreign policy once affected the aggravation of Sino-Australian relations. However, Beijing has also unleashed a powerful Belt and Road Initiative and promises to offer green energy opportunities. This is where Artyom Garin's chapter provides interesting insights into the impact of climate change on these PICs economies. It examines in detail the role of climate agenda in the foreign policy and ODA of the leading actors of the Indo-Pacific, as well as presents the prospects for the rivalry of the PRC and Quad countries in the field of climate change in the South Pacific. If anything, China's military pact with the Solomon Islands in April 2022 has only made China's investment in PICs, a more complicated challenge for the U.S. and its allies especially Australia which has historically been a leading player for PICs.

A country that is extremely vulnerable to climate change is Bangladesh. Hence, the chapter by **Sumy, Sarkar, and Hasan**, titled, *Political Economy of River Ecocide in Bangladesh: A Study in the Context of Dhaleshwari River*, is significant. The authors scrutinize the political-economic factors responsible for the river ecocide that augments the impact of climate change. Ecocide, the deliberate or thoughtless destruction of the environment, has existed throughout the human history. However, it has never been as imminent as it has been in contemporary times. Activities ranging from industrial manufacturing to daily household chores are alarmingly contributing to the phenomenon. Most of the prevailing studies focus on the impact on the environment by industrial activities but do not emphasize the issue of ecocide. This chapter offers a diverse spectrum in addressing river ecocide in Bangladesh through the lens of political economy. The authors seek to explore the impact of ecocide on people's livelihood and socio-environmental sustainability. Findings of the study show that lack of efficient implementation mechanism allows the tannery industry to cause river ecocide although environmental laws and policies are in place. To conclude, the study recommends a responsible, inclusive, and collaborative approach that will tackle the issue of ecocide and prepare Bangladesh to fight climate change constructively.

From the issue of ecocide of a river, the focus shifts to the challenge of understanding the ramifications of space debris. The next chapter titled, *Challenges of Space Debris and Space Drag: Building an International Climate Change Regime* by **Dr. Swasti Rao and Dr. Kunwar Alkendra Pratap Singh** contends that in spite of the growing realization and the development of international regime complex for climate change, humanity has rarely seen any extraordinary measures being taken to tackle the real threats emanating from climate change. This disjunction between theory and praxis is even more glaring in outer space where future power maximization capabilities are unfolding rapidly infesting space with more and more artifacts by both state and non-state parties. The relentless maximization of space capabilities among space faring nations and their corporations, burgeoning space technologies and applications aimed for the use of outer space, have become an increasingly important and even an indispensable factor in everyday life. This chapter analyses two most prominent factors where space faring behavior has complicated the already complex and highly debated issue of climate change. First, the problem of space debris polluting the Lower Earth Orbit (LEO), which, if uncontrolled, can lead to an existential threat for human life. The second factor analyzed in the chapter regards the phenomenon of space drag which is directly related with the greenhouse emissions produced on earth. Key events and patterns that have led to the increase in space debris and space drag are presented to highlight the lack of a strong regime complex in space to control the repercussions of the irresponsible space behavior of countries like China. The authors in conclusion provide suggestions for the strengthening of the Regime Complex for Climate Change by focusing on the concept of Space 2.0 that emphasizes on the techniques of augmentation and reconstitution to make space faring more cost effective, reusable, and transparent and thereby lessening the problem of space cluttering.

Part 2: Institutions and Initiatives

Part 2 comprises a set of chapters analyzing various initiatives intended to mitigate the effects of climate change. In this section, **Kakoli Sengupta's** chapter *Climate Action by the European Union: Making the European Green Deal a Reality* elucidates how Climate Action stands at the very center of the European Union's policies. The EU recognizes the grim reality that climate change and environmental degradation are an existential threat to Europe and the world. The goal of the EU is to make Europe the first climate neutral continent in the world. The European Green Deal set the blueprint for this transformational change. In December 2019, the European Commission presented the European Green Deal committing to climate neutrality by 2050. On July 14, 2021, the European Commission had adopted a set of proposals to make the EU's climate, energy, transport, and taxation policies fit for reducing net GHG emissions by at least 55% by 2030, compared to 1990 levels. Achieving these emissions reductions in the next decade is crucial to Europe for becoming the world's first climate-neutral continent by 2050 and making the European Green Deal a reality.

In order to achieve decarbonization objectives, the European Union leadership believes that emissions must be reduced across all sectors, from industry to transport to farming. The European Green Deal will create new opportunities for innovation and investment and jobs as well as reduce emissions, create jobs and growth, address energy, poverty, and reduce external energy dependency and improve overall health and well-being of European citizens. At the same time, it will ensure opportunities for all, support vulnerable citizens by tackling inequality and strengthening the competitiveness of European companies. The European Commission through the Green Deal also strives to transition toward greener mobility which will offer clean, accessible, and affordable transport even in the most remote areas. The European Commission proposes to increase the binding target of renewable sources in the EU's energy mix to 40%. The proposals promote the uptake of renewable fuels such as hydrogen in industry and transport with additional targets. Besides, the proposals include renovating people's homes and residential buildings for greener lifestyles which will save energy, protect against extremes of heat or cold, and tackle energy poverty. Since nature is an important ally in the fight against climate change, restoring nature and enabling biodiversity offers an efficient and cost-effective solution to absorb and store carbon. To increase absorption of CO_2 and make the environment more resilient to climate change, the Commission proposes to restore Europe's forests, soils, wetlands, and peatlands. The objective of the chapter is to explore the proposals in detail and the scope is to assess the possible and potential impact of the proposals on climate legislation and climate action in the world at large.

An initiative jointly undertaken by France and India, is the International Solar Alliance. **Claudia Astarita and Julius Hulshof**, in their chapter titled, *International Solar Alliance: Testing a New Framework to Approach Energy Shortages*, explain this joint initiative that was launched during the Climate Conference in Paris in December 2015 (COP21). It aims to make an unprecedented effort to promote solar energy to achieve three major goals: creating the conditions for a major decrease in the cost of solar energy; meeting the high energy demand in developing countries; and contributing to the fight against climate change. To achieve these objectives, ISA committed to set new ground rules, norms, and standards for solar energy, regularly discussing with member countries the opportunities of developing innovative capacity-building measures and financial instruments to contribute to the harmonization of public policies, regulations, and prices between the countries. ISA has been conceived as an action-oriented, member-driven, collaborative multilateral platform aimed at reducing users, governments, and investors' uncertainties related to the opportunity of relaunching both solar energy market and infrastructures to enhance energy security and sustainable development. Seven years after ISA's foundation, this chapter discusses both its major achievements as also its failures, as well as to assess the validity and the strength of this multilateral framework as a model for dealing with some of the major energy-related challenges in a post-COVID-19 world. In particular, the chapter discusses the feasibility and the advantages of creating a similar framework to facilitate a multilateral coordination in other areas, such as wind energy and hydro energy.

Much has been written about multilateralism and its significance especially after the Second World War and the risks multilateral institutions face in becoming irrelevant in the present changed milieu (Agarwal 2021). In the next chapter titled, *Multilateralism Efforts in Asia: What's the Way Forward?* **Prathit Singh, Namit Mahajan, and Ritvick Khanna** elucidate how over the past decades, the world has seen persistent attempts by international players to scrutinize a global consensus for acting on climate change, one of the most impending issues of the decade. Beginning at Copenhagen, being ill-sustained at Kyoto, and finding renewal at Paris, the climate change fight has shown us close moments wherein the international community expected a final climate agreement, with predefined cuts on carbon emissions and enhanced investments and pollution control. However, these climate change movements have failed to yield any concise result, only further dividing international parties and altering their approach. The Asian incentive in coming to a global consensus on the climate deal has been a crucial determinant in shaping the narrative and positioning stances at the different climate conferences. The incentive, thus far, has found inadequate actualization due to conflicts of interest between developed and developing nations. The need for a consensus at such a large international stage has contributed to many of its failures owing to the ever-present obsession with multilateralism. Such an approach has only led to a cycle of disruptions with no change whatsoever. Their chapter attempts to explore, primarily from an Asian perspective, the concept of multilateralism which they see as a significant contributing factor resulting in futile agreements. Keeping Asia at the center, the authors explore three questions. First, they examine the ideological flaw in the multilateral obsession that leads to a continuous hindrance of the Asian cause. Then, they analyze the conflict of knowledge and expertise, which Asia and the Global South face while confronting the Global North at the negotiation table. Finally, this chapter investigates various structural flaws in the deliberation process itself which it sees as guided by a multilateral idea of negotiations. Subsequently, this chapter presents the alternatives to the multilateral idea while exploring minilateral methods with a deliberative democratic approach to make climate change agenda more effective.

Part 3: Climate Change from an Indian Perspective

This section comprises four chapters.

Chaitra C. in her chapter titled, *International Climate Governance: Indian Perspectives*, writes how the environmental ethos remain integral to India's civilizational legacies. This according to her has undergirded India's increasing and intense participation in international climate change governance and negotiations. New Delhi has repeatedly repositioned itself from normative to empirical approaches seeking equitable distribution of the carbon space while underlining the historical responsibilities' narrative. Given its emissions profile, economic growth, and leadership role in the developing world, India has explored and benefited from market mechanisms tied to climate adaptation and mitigation. The author contends that India remains committed to global efforts as is demonstrated by India's ambitious initiatives like the International Solar Alliance and Coalition for Disaster Resilient Infrastructure. India's Nationally Determined Contributions (NDCs) to United Nations Framework Convention on Climate Change (UNFCCC) and ratification of Paris Treaty reflect India's engagement with climate change initiatives. Indian Pavilion at the Conference of Parties (COP) had showcased indigenous knowledge and sustainable lifestyles which promises to enrich its soft power as well. At the same time, India needs to address socio-economic development needs of 1.3 billion people, such as access to electricity for all; and also cater to industrial and infrastructural developments. Being the third largest emitter of carbon dioxide, India needs to move away from rhetorical advocacy to appeal to its third world constituencies. While it needs to insist on technology transfer and financial assistance from the developed world, it also needs to partner with them to fast forward its own initiatives.

This chapter examines the evolving contours of Indian position in global climate governance and negotiations by analyzing the approaches India has adopted at various international forums as also in its domestic policies and initiatives. India, for instance, has made major strides in its renewable energy sectors; added 15,000 square kilometer forest cover over the past 7 years. However, in face of exponentially growing energy demand at home, India does not seem in a position as yet to dismantle its fossil fuel dependency. India has to learn from the example of how the recent G7 summit committed to net-zero emissions in the near future has also failed to suspend its own fossil fuels sector. The G7 "nature compact", 30by30 pledge formulated by the U.S. has been endorsed by India. India has joined the Adaptation Action Coalition but is not a part of Carbon Offsetting and Reduction Scheme for International Aviation (CORSIA). With 17.8% of the world's population, India is responsible for a meagre 3.2% of cumulative emissions. From once being in climate denial mode to now emerging as an agenda-setter, India's role in international climate governance is presented in this chapter to be too conspicuous to be marginalized.

Sheeraz Ahmad Alaie takes this analysis and discussion forward as she writes on *Policies as Instruments in Promoting Sustainable Development: Limiting the Climate Change Issues in India*. In her view, climate change is the most critical issue threatening the whole of blue planet. As the most populous, tropical developing country, India has a bigger challenge to cope with the likely consequences of climate change than most other countries that are either small or developed or both. Again, while the climate change is a global phenomenon, its consequences are both local and global. Hence, the role of policies, both at global and country level, is very critical. The Indian climate change policies are featured under both regional and global dimensions with National Action Plan on Climate Change (NAPCC) that was first adopted in 2008 and India's Intended Nationally Determined Commitments (INDC) announced in 2015. This chapter makes an assessment of the role of such national policies in promoting sustainable development and makes an attempt at deciphering the challenges and opportunities within such policy perspectives. It also interrogates the existence of any substantive institutional framework for policy which offers coherence and consistency as to how government should cope with the long-term political challenges of climate change. For India, the critical subject to ponder upon is its characterization in the international climate change landscape. India was seen as a minor contributor to past emissions, but a significant contributor to future emissions. But the logic changes inversely when the same is analyzed on per capita basis. Nevertheless, instead of escaping responsibility, India's national climate policy urgently needs a coherent vision for tackling climate change, that should be reflected in the framing of legislation and policy documents addressing carbon emission-related sectors and aligned with multiple federal levels. This chapter concludes how India could facilitate designing of appropriate institutional frameworks to achieve climate policy objectives of adaptation and mitigation.

Sonia Roy, in her chapter titled, *The Impact of Fridays for Future Movement in Indian Politics*, explains that the "FfF" (Friday for Future) India chapter is an interest group comprising primarily young climate activists and volunteers, who seek to use protest and demonstrations as a means to direct attention of the government and civil society to the alarming effects of climate change. According to the website (fridaysforfuture-india.com) — it is an informal and decentralized association set out to achieve a common goal — *viz.* to protest against global warming, climate change, and perceived government apathy to environmental concerns. The FfF interestingly states that this movement "avoids power-structures", and "is beyond politics" of any kind. But the question the author raises is whether it is possible for a movement to be built on the principle of influencing political developments on climate change by remaining apolitical in nature? While questioning those in power — and their policies — how can this organization define it is being beyond politics? What is the future of such a movement? The particular case of the arrest of Disha Ravi, founder and a young climate change activist of the FfF movement in India by the state, had received worldwide public criticism. Looking into the aggressive stance of the Indian state *vis-à-vis* such climate movements, the chapter seeks to understand the power dynamics at play against these two predominant players — the Indian State on the one hand and the Social Movement Organizations (SMOs) like FfF-India chapter on the other. The hypothesis of this chapter is that despite claiming to be non-political in its objectives, the FfF movement in India has acted increasingly as a lobby in garnering political influence in Indian politics toward its cause.

The final chapter titled, *Emerging Indian Partnerships in Climate Change with Special Reference to COVID-19 Era* is by **Aditi Basu**. The author examines how the cyclones Amphan and Nisarga that had hit India and Bangladesh in May 2021 showcased dire consequences of climate change. Although many countries have taken a lead in deliberating and discussing on various issues of climate change, they had failed to take clear-cut decisions and joint action in reality. India has always viewed climate change as a diplomatic issue, justifying it by saying that the developed countries should probe into this matter because they are the ones who are responsible for causing the problem. Therefore, it is the responsibility of the First World countries to call for global preponderance to address climate change. Although the Global Change Data Lab has declared in 2021 that India is not responsible for the rising temperatures and sea levels, India believes that it needs to carve out a middle path on climate policy that could be accepted both in the domestic and international spheres. It is in the COVID-19 pandemic that India has finally defined and related climate action with climate diplomacy. The extreme climatic conditions like forest fires in the U.S. and Australia, super cyclones in the U.S. and India, locust attacks in South Asia and extreme heatwaves in the U.S., Europe, and Russia have wreaked havoc and led to the destruction of innumerable lives and property. This chapter outlines India's climate agreements with various countries since 2014, i.e., when Prime Minister Narendra Modi took office which was followed by India's participation in the 2015 Paris COP. It also focuses on the pandemic as a trigger for further streamlining India's stand as also increased awareness on climate change. It especially examines India's climate agreements with the U.S. and China during Paris COP to allude to likely future trajectories on how these three largest carbon producing nations can work together.

References

Agarwal, M. (2021). Economic multilateralism in peril. *International Studies*, *58*(4), 425–441. https://doi.org/10.1177/00208817211056741.

Dubash, N. K. (2019). *India in a Warming World: Integrating Climate Change and Development*. Oxford Scholarship online. https://oxford.universitypressscholarship.com/view/10.1093/oso/9780199498734.001.0001/oso-9780199498734-chapter-1.

Chaturvedi, S. and Doyle, T. (2015). *An Introduction: A Critical Geopolitics of 'Climate Fear/Terror': Roots, Routes and Rhetoric.* In Chaturvedi, S. and Doyle, T. eds., *Climate Terror. New Security Challenges Series.* London: Palgrave Macmillan, London. https://doi.org/10.1057/9781137318954_1.

Crossley, É. (2020). Ecological grief generates desire for environmental healing in tourism after COVID-19. *Tourism Geographies, 22*(3), 1–11.

Deutsche Welle (2022). Climate change: Why it is now or never for India. *Made for Minds.* https://www.dw.com/en/climate-change-why-it-is-now-or-never-for-india/a-61000680.

IFPRI (2022). *Global Food Policy Report 2022: Climate Change & Food Systems.* Washington DC: International Food Policy Research Institute. https://ebrary.ifpri.org/utils/getfile/collection/p15738coll2/id/135889/filename/136101.pdf.

IPCC (2022). https://www.ipcc.ch/report/sixth-assessment-report-working-group-3/, accessed on September 9, 2022.

Kedia, S., Pandey, R., and Sinha, R. (2020). Shaping the post-COVID-19 development paradigm in India: Some imperatives for greening the economic recovery. *Millennial Asia, 11*(3), 268–298. https://doi.org/10.1177/0976399620958509; https://journals.sagepub.com/doi/full/10.1177/0976399620958509.

Kelman, I. (2022). https://www.ucl.ac.uk/news/2022/feb/opinion-how-politics-not-climate-change-responsible-disasters-and-conflict.

Koshy, J. (2021). India's percentage COP emissions rose faster than the world average. *The Hindu*, March 1, 2021. https://www.thehindu.com/news/national/indias-percentage-co2-emissions-rose-faster-than-the-world-average/article33965283.ece.

Lu, C. (2022). How climate change fuels hunger. *Foreign Policy*, May 19, 2022. https://foreignpolicy.com/2022/05/19/climate-change-food-insecurity-hunger-drought/.

Mohan, V. (2021). By 2030, put per capita emission to global average: India to G20. *EnergyWorld.com*, July 26. https://energy.economictimes.indiatimes.com/news/renewable/by-2030-cut-per-capita-emission-to-global-average-india-to-g20/84746951.

Mohanty, A. (2020). Preparing India for extreme climate events. *CEEW.* https://www.ceew.in/publications/preparing-india-for-extreme-climate-weather-events, accessed on April 12, 2022.

O'Callaghan-Gordo, C. and Antó, J. M. (2020). COVID-19: The disease of the anthropocene. *Environmental Research, 187*(2020), 109683–109683.

Press Information Bureau, Government of India (2021). National Statement by Prime Minister Shri Narendra Modi at COP26 Summit in Glasgow. Prime Minister's Office, November 1, 2021, https://pib.gov.in/PressReleaseDetail.aspx?PRID=1768712.

Sharma, Y. S. (2022). India can be third largest economy by 2030 on back of four big reforms: Arvind Panagariya. *The Economic Times*, February 25, 2022.

https://economictimes.indiatimes.com/news/economy/indicators/india-can-be-3rd-largest-economy-by-2030-on-back-of-four-big-reforms-arvind-panagariya/articleshow/89834954.cms.

Singh, K. (2019). India's CO2 emissions are growing faster than the US' or China's. *Quarz India*, March 27, 2019. https://qz.com/india/1581665/indias-carbon-emissions-growing-faster-than-us-china-says-iea/.

WHO (2020). https://www.who.int/news-room/fact-sheets/detail/zoonoses, accessed on September 9, 2022.

World Resources Institute (2022). https://www.wri.org/news/statement-ipcc-report-sounds-alarm-climate-clock-ticking-time-act-now, accessed on April 5, 2022.

Section I

Issues

Chapter 2

All Ships Are Not Raised: The Politics of Climate Disasters in the Anthropocene

Saurabh Thakur*

Abstract

The contemporary climate change crisis is fundamentally different from what is classically understood as natural disasters, which refer to events whose impacts are restricted in both time and space. In the Anthropocene epoch, climate change disasters are being interpreted in a new language of crisis and unknown risks; it is riddled with divergent interpretations of climate responsibilities and binary framings of utopian and dystopian future. This chapter argues that the tragedy of the commons on a planetary scale will be hard to avoid unless the ideological and political roots of present-day understanding of disasters are explored and rendered visible. The language of boundaries and limits which dominate the Anthropocene discourse tends to collide with the narratives of

* Saurabh Thakur is an Associate Fellow at the National Maritime Foundation, New Delhi, India. His research sits at the intersection of climate governance, international politics, and sustainability, looking specifically at the climate security and blue economy discourses in the context of South Asia. Currently he holds the CDRI Fellowship (2021–2022) at the Coalition for Disaster Resilient Infrastructure, for which he is working on the project, "Incorporating Infrastructural Resilience in India's Port-led Development Model".

development and global distributive justice, which remain dominant in the Global South. This chapter argues that an ahistorical framing of the Anthropocene presents an analytical challenge to multiple disciplines and especially for the study of International Relations, where the collision of world history with planetary history will be the first to become most visible. The chapter contends that the politics of climate change that seems so far driven by dystopian/utopian futures tends to obfuscate the questions of moral culpability, accountability, and asymmetry of state power and climate justice, which must be the dominant concerns in the present.

Keywords: Anthropocene, Disasters, Paris Agreement, Climate History, Equity, International Relations, Resilience

Introduction

In a rather ambiguous passage, Donald Rumsfeld, the U.S. secretary of defense during the presidency of George Bush Jr., had outlined the policy choices that confronted him during the post 9/11 war on terror. To quote him, "Reports that say that something hasn't happened are always interesting to me, because as we know, there are known knowns; there are things we know we know. We also know there are known unknowns; that is to say we know there are some things we do not know. But there are also unknown unknowns — the ones we don't know we don't know" (Rumsfeld 2002).

The categories he outlined are extremely relevant within the emerging climate change discourse, particularly when one considers the advent of the Anthropocene[1] discourse on climate change. The Anthropocene attempts to trace the roots of the present-day ecological crisis and construct a narrative which holds far reaching implications for an uncertain and unknown future for both humans and environment. The Anthropocene, while it renews the arguments in favor of universalism, so far lacks a common definition. While some point to the Boserupian promises of this new epoch, others are more skeptical about what such promises signify for humanity as a whole (Bonneuil 2015; Jameison 2014; Chakrabarty

---

[1] Anthropocene according to Oxford Dictionary relates to or denotes the current geological age, viewed as the period during which human activity has been the dominant influence on climate and the environment.

2018). Steeped in a language of urgency and apocalyptic change, it proscribes an ambitious socio-political churning in order to secure the future.

President Obama's speech at the Paris COP21 meeting in 2015 conveyed this urgency in a language previously unseen from the reluctant U.S. leadership (Gupta 2016). He warned, "Submerged countries. Abandoned cities. Fields that no longer grow. Political disruptions that trigger new conflict, and even more floods of desperate peoples seeking the sanctuary of nations not their own" (Kirby and O'Mahony 2017:3). The cataclysmic imagery of his speech was a break from the past when political leaders shied away from terming climate change as a disaster. Indeed, in the period since the end of the Cold War, natural disasters have gained more considerable attention due to the rising frequency of extreme weather events and rising greenhouse gas emissions (Rogelj *et al.* 2015; Hollis 2017). This dramatic collapse of natural systems has coincided with the ascendance of the human species as a geological force, acting on par with natural forces (Chakrabarty 2009). The scale of the crisis makes it necessary that the world thinks beyond simple categorization of "dangerous", as enshrined in the UNFCCC text of 1992 Convention (UNFCCC 1992). Moving forward, the Paris Agreement enshrined the two degrees objective and therefore managed to establish a quantitative criterion to access the collective efforts of the parties involved. The argument is also made in favor of going beyond such categorizations that portray how warming beyond 3° should be deemed "catastrophic", and beyond 5° to be "unknown" and so on (Xu and Ramanathan 2017).

The modern understanding of climate change has emerged as a means to think across the scales of space and time and such a project was historically based on a political imperative (Hulme 2004; Drabek 2005; Nordblad 2017; Coen 2018). The science of disaster in the late 19th century emerged with an understanding that natural disasters are not merely a natural catastrophe, but rather they are "always the outcome of an unfortunate conjunction of geophysical circumstances and human choices" (Coen 2018:5). Any attempt to understand a disaster means that both the geographical as well as the historical-social aspects have to be taken into account. Unlike this formulation of disasters, which tends to move back and forth between the natural and the social world, the modern climate prediction modeling has a low spatial and temporal resolution, and it is exclusively focused on the global impacts, often foregoing the local and regional. The Anthropocene discourse runs into a similar roadblock, where the objective reality of planetary scale climate change must confront the subjective and correlate with the localized everyday realities. Disasters are turning into an inescapable part of the experiential reality in the Anthropocene, but it is critical to engage with the affective narratives to understand the relationship between people and climate change and acknowledge that environmental knowledge and world views are inextricably linked with history, politics, and culture.

This chapter confines its analysis in examining threadbare the relevance of disasters in the international politics, especially in the Anthropocene epoch. In the first part, it traces the origins and interpretations of the term like "crisis", "disaster", and "catastrophe", which are a prominent part of the lexicon of the Anthropocene. The second part then traces this new language and its interpretations in the context of international politics. The final section attempts to view the Anthropocene from the perspective of the Global South. This chapter finally contends that the cataclysmic imagination of the climate crisis in the Anthropocene potentially may lead to a post-politicization of climate change debates, wherein the anticipatory modes of governance portend to replace the equitable processes of consensus building with the consensus based on strict scientific expertise and yet offering vague promises of techno-managerial utopia. This foreshadowing of a dystopian future threatens to obfuscate the questions of moral culpability, accountability, asymmetry of state power and justice, which remain the dominant concerns in the present day international politics and everyday life around the world. In the end, this chapter problematizes the universal category of "Anthropos" that perhaps fails to explain the radically segregated futures for the citizens of Global North and Global South even in this new epoch.

Climate Change and Its Interpretations

The normative understanding of climate has been the highlight of modern-day climate change discourses, often exposing the fault lines of debate between the global south and global north, rich and poor, and big versus small states. The rise of environmentalism in the mid-20th century heralded a new era of Climate history, wherein environment protectionism and conservation became the new slogans which shaped the politics and paved the way for the global consensus. Taking a cue from Hardin's *Tragedy of Commons*, Paul Ehrlich's *The Population Bomb* (1968) that predicted overpopulation leading to catastrophic level starvation portrayed this resulting in social upheavals on a disastrous scale. In 1972, the much-debated limits to growth emerged from the Club of Rome Report arguing that the earth is finite and the unlimited growth in population, as well as material goods, would eventually lead to a crisis on the planetary scale (Eastin *et al.* 2011).

While the neo-Malthusian reports like *Limits to Growth* (Meadows *et al.* 1972) and *Revenge of Gaia* (Lovelock 1979) kept the debate centered on issues like overpopulation, food security, migration, and *The End of the World*, a post-colonial critique termed these "Earthrise Era" universalizing hypothesis as imperialist and racist in their outlook toward the Global South (Lazier 2012; Lekan 2014). The developed North focused on efficiency and economics, the states in the Global South took up the banner of Justice and the right to development (Agarwal and Narain 1991). The rise of whole Earth environmentalism in the 1970s drew attention to the planet's ecological limits and carrying capacity. Metaphors of lifeboats were replacing the spaceship earth metaphor for earth as a stable, secure place. The famous blue marble image became a cultural symbol of American environmentalism, exposing the fault lines between the Global North and Global South.

At the first global conference on the human environment (UNCHE) in Stockholm (1972), the "Blue Marble" image was an important motif that marked a Copernican shift in human consciousness. The conference later became famous for the famous speech by the then Prime Minister of India, Indira Gandhi, who articulated the fundamental problem that has since marked the negotiations, its disagreements, and its politics. She argued that many of the environmentalists' arguments about protecting nature were Northern countries' attempts to prevent Southern states from benefiting from development. Poverty was the Southern states' most significant problem; the development paradigm was the solution. She argued:

> The environmental problems of developing countries are not the side effects of excessive industrialisation but reflect the inadequacy of development ... We do not wish to impoverish the environment any further, and yet we cannot for a moment forget the grim poverty of large numbers of people. (Gandhi as quoted in Mathiesen 2014)

In the following decade, along with globalization, the notion of sustainability was to become the most significant paradigm shift in the understanding of the emerging ecological crisis. Swyngedouw (2007:27), however, contests this position by arguing that sustainability attempts to place limits on "the possible" thereby marginalizing any radical socio-economic antagonisms to its approach. This meant that sustainability was interpreted and applied in a variety of ways, most prominently being the ecological modernization discourse, that divides responsibility between corporate markets, and the state, and, "ignoring issues of participation and reducing the rest of society to passive consumers to be provided with enough information to make informed (but market-based) choices" (Gibbs 2000:11).

The first phases of the climate negotiations imagined the climate crisis as a "dangerous" prospect, which could be dealt through global, collective efforts, designed along the lines of the common but differentiated responsibilities (CBDR) and other precautionary principle (UNFCCC 1992). It was a period of constructive ambiguity in the language which was expected to facilitate a diplomatic dialogue on the mitigation of the climate crisis. The period was marked by a sense of incrementalism, which visualized climate change as a structural problem on which there was some scientific and normative consensus (Gupta 2016). By the time of the Copenhagen summit in 2009, the normative consensus had crumbled in the wake of a more concrete scientific consensus from the successive IPCC reports. As the scientific proof of anthropogenic climate change began to grow, the political agreement on the nature of the crisis and division of responsibility for averting any future disasters started to wane.

The post-Copenhagen period provides an interesting case study to understand catastrophism and its impact on the mainstream climate change discourse. The failure of Copenhagen was followed by the unexpected breakthrough at Cancun, where the idea of a two-degree target was mooted, and later adopted in Paris COP21 meeting in 2015. It is also the period when IPCC reports, which generally projected conservative estimates of temperature rises, began to warn about the small window that was available for mitigating global warming to a maximum of 1.5°. Anything beyond a half a degree, scientists argued, would worsen the risks of extreme weather events and widespread poverty. In his address to the conference of parties at Paris in 2015, President Obama called climate change an "existential threat" which portends the fate of our future generations. Describing his visit to Alaska, he spoke:

> I saw the effects of climate change first-hand in our northernmost state, Alaska, where the sea is already swallowing villages and eroding shorelines; where permafrost thaws and the tundra burns; where glaciers are melting at a pace unprecedented in modern times. (Obama, 2015)

The language of crisis, which was the mainstay of the Kyoto Protocol period within international politics, was beginning to be replaced by a more robust and urgent call for action. Much like the central argument of the Anthropocene, the climate crisis was no more a futuristic probability, but an unfolding catastrophe in the present. The COP meetings from 2009 till 2015 rebuilt the UNFCCC framework, by incorporating a range of issues like Loss and Damage, and Climate Justice in a preambular form, whereas the CBDR, which was the operational principle of equity within the negotiations was replaced with and Common But Differentiated Responsibilities and Respective Capabilities (CBDR-RC) according to "evolving national circumstances" (Rajamani 2016). The new regime was universal and ahistorical and futuristic in its outlook. The new pillars and transparency were incorporated within a more flexible regime which was based on intended nationally determined contributions (INDCs) of parties.

Post-Paris World and Fresh Consensus

The near-universal consensus of the Paris Agreement puts forth two contradictory developments: First, is the role of scientific expertise pushing a massive change in the "politics" of climate change discourse. This relationship has remained "inconsistent and apocryphal", and the concept of "best available science" signals a shift within the politics and policymaking where climate science has transformed from advocacy to a "solution- and future-oriented", regulatory science (Beck and Mahony 2018:2). The scientific expertise is increasingly being seen a cogent alternative to political legitimacy. The Paris COP21 also marked the beginning of a polycentric form of governance architecture, whose success depends on the coordinated efforts of states and non-state actors (Devès *et al.* 2018). The scientific community was traditionally assigned the role to experiment with new modes of knowledge and technology, while the application of such new developments was the prerogative of the states (Jasanoff 1990; Barben 2007). This clear boundary was, of course, constantly challenged in the era of globalization with the emergence of institutions like the World Bank and the IMF. The transnational nature of the climate problem in an interconnected world meant that politics and science did not fit into neat compartments, and this interaction extended in multiple dimensions and domains.

The second development, flowing from Paris COP21, is the role of equity and global distributive justice, which has since been reduced to the side-lines of the convention (Thakur 2021). The post-Paris discourse could be described as "a post-equity era of a voluntary and universal climate agreement. In this post-equity world, issues can be addressed by national contributions that will be self-determined" (Klinsky *et al.* 2016:1; Pickering 2012). The new interpretation of CBDR-RC, according to evolving national circumstances marked the end of "differentiation" enshrined in the Kyoto Protocol regime of late 1990s. At the same time, however, even the CBDR-RC remains only in the text of the Paris Agreement as the operationalization of equity remains an unfinished task. These developments are further complicated by the shift toward an INDCs-based approach, where each state now gets to decide its own targets, *vis-à-vis* the other countries. It is interesting to see how international politics experts like Keohane and Oppenheimer (2016) distinguish between legitimacy of the political process and equity and justice issues, thereby making a case that such issues tend to delegitimize the institution and, in the end, prove to be self-defeating from the standpoint of justice itself (Buchanan and Keohane 2006:409).

In the period since the failure of the Copenhagen summit in 2009, the crisis of climate change had been made real through words, speeches, and various agreements on the international front. The vocabulary of crisis, that predominated these interstate negotiations, was either getting replaced by the cataclysmic language or complete denial. The language of moral culpability and differentiated responsibility, which were the hallmarks of the 1992 convention, were losing favor for a more obscure alternative of ambition, which of course remains unstitched so far. Foyle *et al.* (2017:7) point toward the process of acclimatization wherein, "COP21 simultaneously appears as an occasion, for a variety of actors, to lobby for the inclusion of new issues and topics into climate talks" and also "a highly efficient cleansing device where climate change is rendered 'governable' through the deliberate omission of specific issues and alternative approaches to the problem".

The Paris COP21 provided a brief preview of the global environmental governance, where the most vulnerable, marginalized communities emerge at risk of being relocated into a new epoch of Anthropocene, where expertise is being privileged over the day-to-day experiences and struggles of those who sit in these margins. The language of universalism and global innovative solutions has since become ubiquitous both in the speeches of the leaders as well as the texts, yet the real politics of climate change has regressed further back to the domestic sphere. The consensus of Paris COP21 came at the cost of principles of equity and historical responsibility, and yet the outcomes have more or less remained stagnant. The lack of political consensus, which mired the process for two decades, has meant that it is scientific knowledge and its predictive capacities that are now being utilized to shape the future of the climate discourse.

Anthropocene and International Relations

The World Bank Governor Robert Zoellick (2010) makes a case of redundancy of the North–South dichotomy and termed the coinage of "third world" outdated as, he says: "If 1989 saw the end of the 'Second World' with Communism's demise, then 2009 saw the end of what was known as the 'Third World'". Anthropocene discourse centers on these cosmopolitan ideas of planetary security and stewardship, yet the mainstream IR is theorized and practiced on a common-sense logic which is essentially a set of Eurocentric ideas of great power rivalries, norms of sovereignty, and power maximization. The principle of sovereignty, derived from the Eurocentric understanding of the world politics, is at the heart of the Westphalian system and has remained unchallenged since the Second World War and the period of decolonization.

Sovereignty itself is a social construction, as it only came into existence with the treaty of Westphalia in 1648. The nature of sovereignty — as power or responsibility — has been a debated extensively in mainstream IR. Although the Neo-realists and liberal institutional schools differ on the importance of self-interest and collective interests in world politics, they both adhere to the view that State is the central actor which determines all the outcomes in international politics, thereby perpetuating an absolutist, Eurocentric understanding of sovereignty. The UN Charter was only instrumental in further transforming this idea into a universal norm, thereby giving the UN Security Council powers to intervene in the affairs of states in case peace is threatened or breached anywhere by any of the member states. This makes it clear that it is the powerful states that dictate the limits of sovereignty in world politics, and it has managed to survive as a norm within international law because it suits their interests. The universalism embedded in the Anthropocene threatens to overturn this enduring principle of sovereignty in the international politics without acknowledging either its political roots or its post-colonial continuities (Orford *et al.* 2016).

The category of "Third World" or the post-colonial state or the Global South likewise remains relevant as they anchor the political and environmental discourses in the present moment that brings forth the role of empire and powerful states in shaping the international norms. Writing in 1952, a piece titled, "Three Worlds, A Planet" Alfred Sauvy, coined the term "third world", to refer to the underdeveloped, then recently freed countries, which, he feared will go invisible in the midst of cold rivalries, as the obsession with arms race will ultimately trump the concerns of this third world. Alluding to the third estate in the French revolution, he cautioned, "For finally this Third World ignored, exploited, despised as the Third Estate, wants, too, to be something" (Sauvy 1952). This great power dominated international politics continues to remain grounded in such dichotomous representations of the world that propagates binaries like us versus them, civilized versus uncivilized, developed versus underdeveloped, thereby perpetuating the existing asymmetrical relations that are equally visible in the evolution of climate change discourse (Andersson 2012; Akcinaroglu *et al.* 2011).

It is, therefore, important to acknowledge that colonialism was not merely a historical event that happened as the Westphalian system took shape across the world. It has been the very foundation of the international law that the Anthropocene seeks to over cede. The right of self-determination and permanent sovereignty over natural resources holds tremendous significance in the Global South as it is perceived to be a course correction in the path toward its final and real decolonization. Any changes in this world order would require the participation and consent of this underdeveloped world. The new language of utopia/dystopia holds the potential to re-imagine the principle of sovereignty in the anarchical world order and may even provide incentive to the powerful states to rethink their core self-interests (Habib 2015; Comfort 2000). But this language is also suspect to a greater impetus on territoriality and ecological nationalism as is apparent in today's international politics, which is again harking back to the days of hyper-sovereignty.

The UNFCCC regime is a good example of these parallel processes, wherein, at its very inception in 1992, it adopted the principles of sustainable development and Common but Differentiated Responsibilities (CBDR) to form the equitable foundation for any future climate action and placate any fears of environmental colonialism. The subsequent failure of these conventions to translate CBDR principle into legal action, as was the case in the failure of Kyoto Protocol, meant that differential responsibilities have remained voluntary, and at best aspirational, and without any legal consequences. The Paris Agreement of 2015 has therefore pushed the climate action back into the Eurocentric orbit of state-run diplomacy, which is dominated by a language of constructive ambiguity, self-interests, trade-offs, market-based green solutions, and a business-as-usual approach. The tussle between major powers driven international politics and new Anthropocene debate defining contours of climate change, therefore, remains unresolved so far.

The Anthropocene Debate: A View from the South

The climate change discourse grew out of a period in world history when the "global environment" was gaining credence as a political and scientific category (Ingold 2000). The Blue Marble image made famous during the heights of the Cold War, as a symbol of eco-cosmopolitanism, depicted both the infinitude of human progress and the finitude of human endeavors. Incidentally, it is this image that symbolizes both the Whole Earth environmentalism of the 1970s as well as the Anthropocene discourse in the 21st century. This time around the scientific evidence is unequivocally clear on the anthropogenic causes of climate change disaster. Lekan (2014:197) makes clear that the post-colonial critique of the Anthropocene is, "not a call back to a local 'ethics of proximity', as a place of resistance to the Whole Earth, nor does its 'hermeneutics of suspicion' toward discourses that naturalize race, class, or gender miss the contradictory registers of human agency in the Anthropocene". As made clear in the writings of Chakrabarty (2009) the Anthropocene poses the most significant challenge to post-colonial suspicion of the Universal framing. He claims, "in an age when the forces of globalization intersect with those of global warming, the idea of the human needs to be stretched beyond where postcolonial thought advanced it" (Chakrabarty 2012:15).

We live in an age of reflexive modernity, where societies across the world are susceptible to unknown, unforeseen risks (Beck & Wehling 2013). This fear of the unknown transforms the complex notions of basic human needs and well-being into a simple hedging against the future risks. Evans and Reid (2013:83) reflect upon this proposition and ask: What does it mean to live dangerously? No longer, we are told, should we think in terms of evading the possibility of traumatic experiences. For catastrophic events are not just inevitable but also learning experiences from which we have to grow and prosper, collectively and individually. Vulnerability to the threat, injury, and loss have to be accepted as a reality of human existence. This line of argument finds an alarming resonance in the 19th century debates on famine. Dissecting the genesis of famines, scholars such as Sen (1981), came to conclusion that, such disasters were an outcome of food shortage which are equally a result of natural factors and the failure of the socio-political systems. Davis's critical inquiry into this subject in *Late Victorian Holocausts: El Niño Famines and the Making of the Third World* argues that famines were not so much the result of the failure of the British empire but an outcome of the meticulous decisions of "rajas and viceroys" which turned droughts into famines. He terms these late Victorian famines, which were a result of an exploitative imperial economics and environmental instability, as a "wars over the right to existence" (Davis 2002). Therefore, there is bound to be suspicions of the part of post-colonial states who have long viewed universalism as a fundamental aspect of the colonial domination and imperial control (Ashcroft *et al.* 1995:55). The meta narrative of Anthropocene ignores the social heterogeneity of the present world and thereby marginalizes and suppresses the voices, social practices, and conventions that do not conform with the category that is being raised to the universal.

Although such a suspicion has legitimate grounds in colonial history, the planetary scale of the climate change crisis clearly dictates that some universal categories cannot be outrightly rejected. The material reality of the Anthropocene and its consequences will be universally shared. Therefore, it creates an ethical imperative to radically transform human societies. Universality is a critical concept which tends to raise the particular to the category of "universal". Often the group which holds the material advantage projects its values and concerns as universal, thereby rendering all dissent and alternatives as ineffectual and unthinkable. The climate disaster moves beyond this simple representation of universality, as it is already a material reality, established through scientific facts. The dialectic between the "universal" and the "particular" once again play out in the Anthropocene epoch, and the task that confronts human civilization is to avoid building the climate catastrophe in absolute terms. This absolutist thinking is particularly prevalent in the construction of the Anthropocene future in the language of utopias and dystopias; the techno-managerial utopias, where "Mankind" has finally conquered "Nature".

The death of "Nature" in the aforementioned utopias or dystopias is celebrated, and the future is viewed as, "as a promise, a lure and a temptation" (Popper as cited in Raskin 2002:1). The eschatological visions of dystopic future which is made ubiquitous in the cultural realm through disaster movies and science fiction novels also shape and gets shaped by politics. In either of the scenarios, the universal category of "Anthropos" refers to a collected mass of humanity which gets stripped of its diversity, antagonisms, and differences (Tsing 2005). It is an ahistorical monolith which stands on the edifice of a linear time, which views human history to a set of technological and material advances. It is bereft of any stories of the indigenous struggles, cultural diversities, or the history of diversity, conflict, culture, extra-social relations, and colonial brutalities (Malm and Hornborg 2014). It renders all human–nature relationships, and therefore politics, irrelevant and invisible, and technology and risk managers get broadcasted as the new seers of humanity. Such a reductionist view of history turns technology, more specifically greener technology, and greener markets into the new keys to a yet unseen future (Stubblefield 2018).

This technological fetishism of the Anthropocene, with its faith in yet unseen technologies of the future and its gross capacity to quantify and profit from all future risk, is bound to fall prey to the Promethean gap, which refers to the deep chasm between, "what humans can produce with the help of technologies and the capacity of imagining the adverse effects these technologies can have" (Anders as cited in Fuchs 2017:582) Hiroshima provides one such crucial representation of such 20th century nuclear catastrophism. The War on Terror was an exercise in securitization that normalized a crisis and produced a situation, where repealing the emergency measures became unthinkable (Fisher 2009). The "war on climate" conjures up similar landscape of metaphors which create a fertile ground for the creation of both material as well as discursive realities of borders, walls, and fences. It is a world armored and closed in anticipation of the other, the climate migrant, the refugee. In most cases, these limits are being imposed by a "minority world", which increasingly fears influx of the "majority world", which itself is failing to adapt to the rising temperatures (Chaturvedi and Doyle 2015:110).

Rosenthal (1998:160) outlines the three main factors which undergird our understanding of disasters. The transnational nature of the disasters, mediazation, or the role of media is shaping the subjective notion of what constitutes a disaster, and *politicization*, which signify the language used to describe a disaster. These three factors play a key role in shaping how a disaster is framed and acted upon. The Anthropocene adds another layer to these factors. We are now confronted with a politics of anticipation which lays greater faith in the post-political solutions as the new panacea, thereby ignoring the post-colonial realities of the global south (Beck and Mahony 2017; Groves 2017; Granjou 2017; Jónsdóttir 2012). The post-political here does not signify the end of politics but instead points to the process whereby the realm of the political is curtailed and shredded to simple techno-managerial questions of efficiency and consensus. It is a process of shutting down any competing claims of existing asymmetry in power relations (Paddison 2009:5). This discourse diverts attention from the notions of responsibility and accountability in the present and its ahistorical approach to the crisis renders the quest of global distributive justice impossible.

Indeed, guided by politics, the language of "catastrophe" tends to conflate probability with possibility, leading to a form of anticipatory governance, imbued in the logic of resilience, where the life of the subject transforms into a series of dangerous events (Evans and Reid 2013). Therefore, it is critical to problematize the major-powers-driven universalism and catastrophism that are built into current Anthropocene discourse. The critical challenge remains that the modern-day regimes of big data accounting and "disaster capitalism" tend to bet on disasters in this new economy of catastrophe (Appadurai 2013; Klein 2007). The modern-day sites of disasters are, in fact, far removed from the traditional understanding of disasters, which performed a function of social leveling and collective action among communities. These modern-day "Heterodystopias" are, "spaces managed as and in anticipation of a world of a dystopian climate crisis that is at once stages for future interventions and present-day spectacles of climate security" (Cons 2018:266; Foucault 1998). Naomi Klein sums up the disasters in the Anthropocene in following words:

> Today they are moments when we are hurled further apart, when we lurch into a radically segregated future where some of us will fall off the map, and others ascend to a parallel privatised state, one equipped with well-paved highways and skyways, safe bridges, boutique charter schools, fast-lane airport, terminals, and deluxe subways. (Klein 2007:48)

The scientific evidence for all this is clearly indicative; whether one captures the enormity of this moment in the term Anthropocene or any other, of a future — which will be disruptive — holds no recorded historic precedent to rely upon for our conclusions. "...[a]s the Anthropocene becomes ascendant both analytically and politically, it becomes vital to question its imaginary, how it constructs nature and Humanity, how it influences and constrains responses to ecological crises, and what the long-term implications of operating within this imaginary are" (Stubblefield 2018:1). The enormity of any planetary-scale disaster, with roots in deep history and various inflection points throughout world history, brings forth the commonality of the problem and it ends up creating a universal category of "Anthropos" who remains ahistorical and under-analyzed so far.

Conclusion

The project of conservationism, which evolved into the environmentalism as we know today, was once a part of the civilizational project of the West, wherein the colonial subject and their territory was treated as the "living laboratories" for new forms of social, economic, and ecological experiments (Tilley 2011). The suspicion of the universal in the Global South emerges from this history of racism and colonialism. The post-industrial states' desire for global regulations to stem the ill-effects of industrial development runs into the principles of self-determination, which the southern states dutifully guard as they climb the ladders of economic independence. Any discourse which resembles the language of coercion and displaces all other modes of comprehending and dealing with the natural disasters will eventually prove ineffectual in dealing with egalitarian ideas of equity and social equity, which remain relevant in the explicating the global South discourse.

In this ongoing churn of narratives, the Anthropocene offers a new framing of the climate change driven crisis, where disasters are not a "punctual moment" which awaits us in the future, but rather forms part of the gradually unraveling present and what caused such a situation to arise can be traced back to world history (Fisher 2009). Nevertheless, the human existence in this new epoch will be rife with deep anxieties and forecasts about the future, and therefore it is critical to explore this future through the lens of equity and redressal of systemic injustices through what Appadurai (2013:295) describes as the ethics of possibility which widens the horizons of hope and imagination and increases the capacities to aspire, thereby building a more creative and inclusive citizenship. The Blue Marble image gives humanity an illusion that it sits outside of the environment, when in fact, "the environment cannot be accurately understood as an object of governance because exiting it is as much a conceptual impossibility as a scientific one" (Natarajan and Dehm 2019).

The tragedy of commons on a planetary scale will be hard to avoid unless the ideological roots of the hidden present-day "truths" are explored and rendered visible. The dystopian major-powers' driven framing of disasters in the future holds the risk of extrapolating the dismal present and closing the possibility of imagining any alternative futures. Historically, there have been many occasions when the process of production of new forms of knowledge contributed toward prompting epistemic decolonization, which subverted the colonial practices in unpredictable and unintended ways (Tilley 2011). Disasters are inherently linked with the unresolved questions of development and the Anthropocene discourse holds the potential to expose this relationship between the natural and social worlds. The Anthropocene, with a push for rigorous scientific evidence and a collective human consciousness, therefore, promises to unravel critical new modes of interconnections and democratizing the knowledge creation process by bringing forth the concerns of the marginalized.

References

Agarwal, A. and Narain, S. (1991). *Global Warming in an Unequal World: A Case of Environmental Colonialism*. Centre for Science and Environment.

Akcinaroglu, S., Dicicco, J. M., and Radziszewski, E. (2011). Avalanches and olive branches: A multi-method analysis of disasters and peace-making in interstate rivalries. *Political Research Quarterly, 64*(2), 260–275.

Andersson, J. (2012). The great future debate and the struggle for the world. *The American Historical Review, 117*(5), 1411–1430.

Appadurai, A. (2013). *The Future as Cultural Fact. Essays on the Global Condition*. London: Verso Books.

Ashcroft, B., Griffiths, G., and Triffin, H. eds. (1995). *The Postcolonial Studies Reader*. London: Routledge.

Barben, D. (2007). Changing regimes of science and politics: Comparative and transnational perspectives for a world in transition. *Science and Public Policy, 34*(1), 55–69.

Beck, U. and Wehling, P. (2013). The politics of non-knowing: An emerging area of social and political conflict in reflexive modernity. *The Politics of Knowledge*, 33–57, Routledge.

Beck, S. and Mahony, M. (2017). The IPCC and the politics of anticipation. *Nature Climate Change*, 7(5), 311.

Beck, S. and Mahony, M. (2018). The politics of anticipation: The IPCC and the negative emissions technologies experience. *Global Sustainability*, 1, e8.

Bonneuil, C. (2015). The geological turn: Narratives of the Anthropocene. In Hamilton, C., Gemenne, F. and Bonneuil, C. eds., *The Anthropocene and the Global Environmental Crisis* (pp. 17–31). London: Routledge.

Buchanan, A. and Keohane, R. O. (2006). The legitimacy of global governance institutions. *Ethics & International Affairs*, 20(4), 405–437.

Chakrabarty, D. (2009). The climate of history. Four theses. *Critical Inquiry*, 35(2), 197–222.

Chakrabarty, D. (2012). Postcolonial studies and the challenge of climate change. *New Literary History*, 43(1), 14–15.

Chakrabarty, D. (2018). Anthropocene time. *History and Theory*, 57(1), 5–32.

Chaturvedi, S. and Doyle, T. (2015). *Climate Terror: A Critical Geopolitics of Climate Change*. Basingstoke: Palgrave Macmillan.

Coen, D. R. (2018). *Climate in Motion*. Chicago: University of Chicago Press.

Comfort, L. K. (2000). Disaster: Agent of diplomacy or change in international affairs? *Cambridge Review of International Affairs*, 14(1), 277–294.

Cons, J. (2018). Staging climate security: Resilience and heterodystopia in the Bangladesh borderlands. *Cultural Anthropology*, 33(2), 266–294.

Davis, M. (2002). *Late Victorian Holocausts: El Niño Famines and the Making of the Third World*. London: Verso Books.

Devès, M. H., Lang, M., Bourrelier, P. H., and Valérian, F. (2018). Rethinking IPCC expertise from a multi-actor perspective. In Serrao-Neuman, S., Coudrain, A., and Coulter, L. eds. *Communicating Climate Change Information for Decision-Making*. Berlin: Springer, pp. 49–63.

Drabek, T. E. (2005). Sociology, disasters and emergency management: History, contributions, and future agenda. In McEntire, D. ed. *Disciplines, Disasters and Emergency Management: The Convergence of Concepts Issues and Trends from the Research Literature*. Springfield: Charles C. Thomas Pub. Ltd.

Eastin, J., Grundmann, R., and Prakash, A. (2011). The two limits debate: Limits to growth and climate change. *Futures*, 43(1), 16–26.

Ehrlich, P. R. (1978). *The Population Bomb*. New York: Ballantine Books.

Evans, B. and Reid, J. (2013). Dangerously exposed: The life and death of the resilient subject. *Resilience*, 1(2), 83–98.

Foucault, M. (1998). Different Spaces. Translated by Robert Hurley. In *Essential Works of Foucault*, 1954–1984, Volume 2: Aesthetics, Methods, and

Epistemology, edited by James D. Faubion, 175–86. New York: New Press. Originally published in 1984.

Fisher, M. (2009). *Capitalist Realism: Is There No Alternative?* Winchester: John Hunt Publishing.

Fuchs, C. (2017). Günther Anders' undiscovered critical theory of technology in the age of big data capitalism. *Triple C: Communication, Capitalism & Critique*, *15*(2), 582–611.

Gibbs, D. (2000). Ecological modernisation, regional economic development and regional development agencies. *Geoforum*, *31*(1), 9–19.

Granjou, C., Walker, J., and Salazar, J. F. (2017). The politics of anticipation: On knowing and governing environmental futures. *Futures*, *92*, 5–11.

Groves, C. (2017). Emptying the future: On the environmental politics of anticipation. *Futures*, *92*, 29–38.

Gupta, J. (2016). Climate change governance: History, future, and triple-loop learning? *Wiley Interdisciplinary Reviews: Climate Change*, *7*(2), 192–210.

Habib, B. (2015). Climate change, security and regime formation in East Asia. In Pacheco Pardo, R. and Reeves, J. eds., *Non-Traditional Security in East Asia: A Regime Approach*. London: Imperial College Press, pp. 49–72.

Hollis, S. (2018). Bridging international relations and disaster studies: The case of disaster-conflict scholarship. *Disasters*, *42*(1), 19–40.

Malm, A. and Hornborg, A. (2014). The geology of mankind? A critique of the Anthropocene narrative. *The Anthropocene Review*, *1*(1), 62–69.

Hulme, M. (2009). *Why We Disagree About Climate Change: Understanding Controversy, Inaction and Opportunity*. Cambridge: Cambridge University Press.

Ingold, T. (2003). Globes and spheres: The topology of environmentalism. In Milton, K. eds., *Environmentalism: The View from Anthropology*. London: Routledge, pp. 39–50.

Jamieson, T. (2014). Shelter from the storm: Natural disasters and international relations theory. *APSA 2014 Annual Meeting Paper SSRN*.

Jasanoff, S. (1990). *The Fifth Branch: Science Advisers as Policy*. Harvard: Harvard University Press.

Jónsdóttir, Á. (2012). Scaling climate: The politics of anticipation. In Hastrup, K. and Skrydstrup, M. eds., *The Social Life of Climate Change Models*. London: Routledge, pp. 138–153.

Keohane, R. O. and Oppenheimer, M. (2016). Paris: Beyond the climate dead end through pledge and review? *Politics and Governance*, *4*(3), 142–151.

Kirby, P. and O'Mahony, T. (2017). *The Political Economy of the Low-Carbon Transition: Pathways Beyond Techno-Optimism*. Berlin: Springer.

Klein, N. (2007). Disaster capitalism: The new economy of catastrophe. *Harper's (October)*, 47–58. https://harpers.org/archive/2007/10/disaster-capitalism/.

Klinsky, S., Roberts, T., Huq, S., Okereke, C., Newell, P., Dauvergne, P., and Keck, M. (2017). Why equity is fundamental in climate change policy research. *Global Environmental Change*, *44*, 170–173.

Lazier, B. (2011). Earthrise; or, the globalization of the world picture. *The American Historical Review, 116*(3), 602–630.

Lekan, T. M. (2014). Fractal Earth: Visualizing the global environment in the Anthropocene. *Environmental Humanities, 5*(1), 171–201.

Lovelock, J. (2006). *The Revenge of Gaia: Earth's Climate in Crisis and the Fate of Humanity.* New York: Basic Books.

Mathiesen, K. (2014). Climate change and poverty: Why Indira Gandhi's speech matters. *The Guardian.* https://www.theguardian.com/global-development-professionals-network/2014/may/06/indira-gandhi-india-climate-change.

Meadows, D. H., Meadows, D. H., Randers, J., and Behrens III, W. W. (1972). *The Limits to Growth: A Report to the Club of Rome.* Potomac Associates.

Natarajan, U. and Dehm, J. (2019). Where is the environment? Locating nature in international law. *TWAILR: Third World Approaches to International Law Review.*

Nordblad, J. (2017). Time for politics: How a conceptual history of forests can help us politicize the long term. *European Journal of Social Theory, 20*(1), 164–182.

Obama, B. (2015). COP21, Session 1 Address. delivered November 30, 2015, Le Bourget, Paris.

Orford, A., Hoffmann, F., and Clark, M. eds. (2016). *The Oxford Handbook of the Theory of International Law.* Oxford: Oxford University Press.

Paddison, R. (2009). Some reflections on the limitations to public participation in the post-political city. *L'Espace Politique [En ligne], 8.*

Pickering, J., Vanderheiden, S., and Miller, S. (2012). "If equity's in, we're out": Scope for fairness in the next global climate agreement. *Ethics & International Affairs, 26*(4), 423–443.

Rajamani, L. (2016). Ambition and differentiation in the 2015 Paris Agreement: Interpretative possibilities and underlying politics. *International & Comparative Law Quarterly, 65*(2), 493–514.

Raskin, P., Banuri, T., Gallopin, G., Gutman, P., Hammond, A., Kates, R., and Swart, R. (2002). *Great Transition: The Promise and Lure of the Times Ahead.* Boston: Stockholm Environmental Institute.

Rogelj, J., Luderer, G., Pietzcker, R. C., Kriegler, E., Schaeffer, M., Krey, V., and Riahi, K. (2015). Energy system transformations for limiting end-of-century warming to below 1.5 C. *Nature Climate Change, 5*(6), 519.

Rosenthal, U. (2005). Future disasters, future definitions. In Perry, R. W. and Quarantelli, E. L., *What is a Disaster?* London: Routledge, pp. 165–178.

Rumsfeld, D. (2002). Secretary Rumsfeld press conference at NATO headquarters, Brussels, Belgium, June 6. https://www.nato.int/docu/speech/2002/s020606g.html, accessed on October 18, 2019.

Sauvy, A. (1952). Three Worlds, One Planet. *L'observateur,* p. 14.

Sen, A. (1981). *Poverty and Famines: An Essay on Entitlement and Deprivation.* Oxford: Clarendon Press.

Stubblefield, C. (2018). Managing the planet: The Anthropocene, good stewardship, and the empty promise of a solution to ecological crisis. *Societies, 8*(2), 38.

Swyngedouw, E. (2007). The post-political city in BAVO GB. In Pauwels, M. ed.
Urban Politics Now, Re-Imagining Democracy in the Neoliberal City.
Rotterdam: Netherlands Architecture Institute Publishers.

Thakur, S. (2021). From Kyoto to Paris and beyond: The emerging politics of
climate change. *India Quarterly*, 77(3), 366–383. doi:10.1177/
09749284211027252.

Tilley, H. (2011). *Africa as a Living Laboratory: Empire, Development, and the
Problem of Scientific Knowledge, 1870–1950*. Chicago: University of
Chicago Press.

Tsing, A. L. (2005). *Friction: An Ethnography of Global Connection*. Princeton,
N.J.: Princeton University Press.

UNFCCC (1992). *United Nations Framework Convention On Climate Change.*
United Nations, FCCC/INFORMAL/84 GE. 05-62220 (E) 200705,
Secretariat of the United Nations Framework Convention on Climate
Change, Bonn, Germany, 24 pp. unfccc.int/resource/docs/convkp/conveng.
pdf.

Xu, Y. and Ramanathan, V. (2017). Well below 2 C: Mitigation strategies for
avoiding dangerous to catastrophic climate changes. *Proceedings of the
National Academy of Sciences, 114*(39), 10315–10323.

Zoellick, Robert B. (2010). *The End of the Third World? Modernizing
Multilateralism for a Multipolar World*, delivered at the Woodrow Wilson
Center for International Scholars, Washington, DC, April 14, 2010; World
Bank, Washington, DC. © World Bank. https://openknowledge.worldbank.
org/handle/10986/29639 License: CC BY 3.0 IGO.

Chapter 3

Climatic Politics of Non-State Actors in the Post-Pandemic Era: Insights from Eco-Cinema of the Global South

Anand Sreekumar*

Abstract

The past decades, culminating in the Paris Agreement of 2015 and the coronavirus pandemic since 2020, have been shaped by myriad manifestations of the politics of climate change. While the sites of politics have rightly percolated from the high politics of states to the low politics of non-state actors, the literature on the latter is overwhelmingly dominated by civil society, grassroots movements, etc. In this backdrop, it is unfortunate that the politics of non-state actors like filmmakers has been largely overlooked and this is especially true of their interventions in climate change discourse. In the post-literate age where visual media has emerged as the supreme source of information, eco-cinema has emerged as a critical site of representation of climate change. Backed by a theoretical framework of visibility strategy, this chapter deploys the methodology of a comparative visual content and discourse analysis to juxtapose eco-cinema from the Global North (*Corals and Our Planet*) and Global South (*Kadvi Hawa, Nelson, The Weeping Apple Tree*, etc.). It makes an

---

*Author is, Research Scholar at Centre for International Politics, Organization and Disarmament, School of International Studies, Jawaharlal Nehru University, New Delhi.

attempt to examine a whole spectrum of visual media, from documentaries to movies, representing diverse experiences of climate change across the world. Based on this comprehensive analysis, it contends how the politics of filmmakers in the Global North is largely shaped by "pre-trauma" anticipating a future dystopia, fetishization of charismatic megafauna, subliminal landscapes, etc., expressed through a cinematic language of excess. However their counterparts from the Global South, while sharing important convergences like trauma/loss, diverge on the question of the temporality of trauma as situated in the mundane every day, in the backdrop of socio-political injustices, echoing a post-disaster realism. Further they signify more nuanced stories of suffering, adaptation, and resilience mirroring the daily realities of the most marginalized people who are the most affected as well as deploy innovative and unconventional aesthetic registers. This chapter in the end concludes arguing that there is a need to look beyond hegemonic modes of representation in the post-pandemic era and how it is signified by the increasing incidences of climate-change disasters as well as our enhanced vulnerability to them.

Keywords: Eco-cinema, Environmentalist Documentaries, Climate Change, Visibility Strategy, Non-State Actors

Introduction

From the latter half of the 20th century, there have been myriad debates surrounding climate change. The rise of consumption of fossil fuels on an unprecedented scale, large-scale deforestation, rampant industrialization, etc., have led to a dramatic shift in weather patterns and climatic variations on a global scale. In fact, man-made climate change has been so distinct and marked that even on a historical geological scale the novel term "Anthropocene" has come to be widely used to describe the current era. In this, international politics and the discipline of international relations have not been indifferent to these critical questions of climate change discourse. The end of the Cold War — taking the focus off the nuclear Armageddon — has led to securitization of the climate change along with other transnational concerns like terrorism, pandemics, etc. To the least, the transnational scale of the problem of climate change has necessitated a transnational response. This has translated into myriad international regimes and institutions (both formal and informal) to tackle climate change challenge.

The most prominent interventions of recent times in this regard range from the Rio Earth Summit of 1992 and the genesis of the UNFCC through the Kyoto conference of 1997, the deadlocks of Conference of Parties at Copenhagen in 2009 and Cancun of 2010 to the "resolution" in the form of 2015 Paris Agreement. These past decades have also seen constantly evolving national positions ranging from climate-skepticism to deep ecologism manifested in the form of myriad eco-political approaches of state entities like the U.S., China, EU, Russia, India, etc. What is significant in these interventions is the prominent role of the actors constituting the so-called "high politics". The majority of the debates on environmental politics has revolved around state actors, their positions, and diplomatic interventions. This, however, dwarfs the agency and the role of non-state actors in both examining as also redressing the questions of climate change.

From the rise of the New Left in 1960s and global environmentalism of 1970s to the increased prominence of the likes of Greta Thunberg, there is now increasing recognition of the role played by non-state actors that include civil society actors, grass-root movements, and so on. However, even as one moves beyond the "high" spaces of states, there is still an under-appreciation of the eco-politics which plays out in the "low" spaces of cinema or pop culture. Despite a significant proliferation in eco-cinema ranging from movies and cli-fi to documentaries and short-films, the agency and the politics of environmental films remain relatively under-explored, especially those from the Global South. What brings relief is that increasingly there have been scholarly interventions in disciplines like film studies, rhetorical studies, culture studies, etc., which have tackled the politics of eco-cinema through different lenses. And, it is in this backdrop that in order to strengthen the integration of IR with these non-state interventions, this chapter aspires to add to the canon of this literature. It does so by analyzing the two recent climate change documentaries and how they reflect, if not reinforce, the dominant representations of climate change through a very specific form of aesthetic treatment.

It begins by making an argument that all these interventions have their political limitations; seeking to substantiate this proposition by juxtaposing it with four transmedial interventions from India. These represent the eco-politics of four separate distinct regions impacted by climate change in their respective manners. These interventions depict non-hegemonic manners of representing climate change, resulting in unconventional and innovative forms of aesthetic registers and embodiment, besides highlighting socio-political and environmental discourses which are prominently missing or understated in most of the dominant forms of cinematic representation of climate change. After setting the stage for this aesthetic turn in IR which empowered scholars to examine the "low data" of pop-culture including films, it highlights the theoretical and methodological framework adopted. It examines eco-cinema of the Global North using two Netflix environmental documentaries and how they reinforce the existing dominant eco-political frames. In the succeeding sections, it juxtaposes this with a few transmedial eco-cinematic interventions of the Global South, exploring if they represent an alternative nuanced (and perhaps more effective) reality of climate change using unconventional and innovative aesthetic registers. There is also a nuanced representation of the realities of climate change by showcasing the pain and the ambiguities involving the realities of adaptation among the most vulnerable communities. It concludes in the post-pandemic era which has seen both crippling economic loss as well as climate change disasters across both the Global North and South; being a wakeup call for looking beyond hegemonic forms of representations while it also highlights various limitations of the emancipatory potential of such representations.

The Aesthetic Turn in International Relations

The aesthetic turn in general can be described as a multidisciplinary critical attitude which informed IR in the latter part of the 20th century. Drawing from disciplines as diverse as feminism, sociology, psychology, culture studies, etc., it was illustrative of a wider shift in IR theory since the end of the Cold War. A critical emancipatory attitude intertwined with a focus to move beyond the echelons of high politics of IR, coupled with the end of isolationism in IR led to a liberal influx of theories from other disciplines especially from the 1990s. There was thus an evolving acknowledgment of the need to expand the disciplinary borders of IR (Åhäll 2008; Carter and Dodds 2014:1,1–13).

This expansion not only led to the emergence of substantial scholarship on the evolving intersections of international politics and art, pop culture, etc., it was also necessitated by the widespread recognition of the realities of a post-literate global society whereby audio-visual media has emerged as the major source of information. Once relegated to the realm of low or micro-politics, there has emerged an increasing recognition that art, pop-culture, movies, etc., are far more significant than mere second-order representations. Some scholars even went further to argue that one should move beyond a positivist cause–effect relationship to conceptualize pop culture and international politics as a continuum. This implies that they share a symbiotic relationship, whereby they are inextricable, intimate, and even intertextual (Grayson, Davies, and Philpott 2009; Wang 2013:32,156).

This has had significant implications on the study of these intersections between IR and various forms of pop culture. For instance, the status of movies and films were elevated from that of a mere entertainment or mimetic representational medium to an intrinsic component of international politics. Scholars have sought to flesh out and conceptualize the variegated relationship between movies and international politics. Carter and Dodds (2014:4) have shown the importance of movies in critical geopolitics, highlighting the examples of how movies constitute cartographies and spaces as well as unsettle and challenge them. Movies are performative in that they constitute the political worlds and also influence political events and crises ranging from George Bush's Top Gun moment to the reaction of Kazakhstan to Borat (Carter and Dodds 2014:108).

There are multiple prisms to analyze these interfaces between movies and international relations. They can be categorized in terms of (a) the global political economies of movie production, diffusion, and reception, (b) the forms of the international "flows" of pop culture and the barriers to the flows, (c) the representation of norms, identities, spaces, and political events and intertextuality, and finally (d) the politics of the consumption of pop culture (Hozic 2014; Weldes and Rowley 2015:13–21,229). Of these four, this chapter primarily concerns itself with the third aspect of the representation within cinema. Cinema plays a significant role in perpetuating global "mythologies" which signify ideological relationships between the world and humanity. Cinema is thus an ideological machine whereby ideological battle lines are drawn. It can even be termed as a rhetorical medium "addressing, constituting and mobilising people" (Parson 2021:13). Movies reflect, mirror, influence, and even reinforce the dominant forms of societal processes (Parson 2021:10–11). While there are movies which justify and uphold existing international power structures, there exists oppositional cinema which challenges their hegemony, even propounding alternative, emancipatory visions (Shapiro 2009; Der Derian 2010:1,19).

Another distinct yet interrelated aspect includes the possibilities of cinema *vis-á-vis* affect, emotions, and feelings. Cinema is a multisensory medium, involving and interacting with the body, both of the aesthetic subjects depicted as well as that of the viewer (Parson 2021:1011). Cinema thus not only reflects the dominant configurations of affect at a particular spatial-temporal location but also constitutes structures of feeling both at an individual and collective scale (Leyda *et al.* 2016:11). In a sense, cinema reflects the deep-seated anxieties and fantasies associated with a particular socio-historical epoch. This explains why movies and cinema have explored multiple global and international events. There have been studies on how cinema represents the war on terror, the complexities of the Cold War, the migration debates with reference to EU, etc. (Dodds and Hochscherf 2020; Carter and Dodds 2014; Shapiro 2009:18,43,108).

Apart from these specific events and episodes, there also exist studies on how cinema normalizes or influences certain modes of being or forms of order in the international arena, e.g., glorification or vilification of war, states and borders (Der Derian 2010:108). There are also interesting studies highlighting the intersection of movies, documentaries, affect, and everyday international relations (Callahan 2015:891). Having broadly defined the contours of the aesthetic turn, thereby contextualizing my study on the representations of climate change, next section focuses on the theoretical and methodological framework employed in this examination of eco-cinema as representative of non-state actors especially from perspectives of the Global South.

Theoretical and Methodological Framework

Gillian Rose employs a four-fold typology in broadly categorizing the sites of study of visual media into its production, the very site of the media, its diffusion, and finally its reception. This chapter concerns itself with the site of the film in terms of how it represents what it does. At this limited site of the visual media, it examines certain aspects of the content, form, and discourses which are being propagated (Rose 2016:198). In a sense, this study employs the visibility strategy whereby one studies the frames of the movie under examination (Callahan, 2020:1). It then highlights the prominent themes which are most prominently represented as well as those aspects constantly marginalized or relegated to the backdrop or even overlooked altogether. In other words, it critically examines the frames to highlight both the visible, as also the invisible. This therefore seeks to inform the theoretical basis of the critical analysis of both exclusions and inclusions.

This chapter, more specifically, adopts the approach of eco-criticism. Eco-criticism is rooted in the new materialism which emerged in the latter half of the 20th century which sought to critique the anthropocentric assumption inherent in most of social and political thought and to extend the agency to entities beyond human beings. There exist myriad manifestations of eco-criticism including environmentalism, deep ecology, eco-feminism, etc. In other words, the site of the eco-cinema is considered as an analytical text where this chapter attempts a critical examination of the authenticity of the forms of environment and nature being represented on screen, thereby deconstructing various aspects of the environmentalist discourse which are highlighted as also those which are marginalized (Parson 2021). This chapter thus adopts thinking ecologically as a method in attempting a combination of discourse analysis and ideological criticism.

Climate Change and International Relations

Barring certain exceptions, IR as a discipline has relatively under-explored the interface of climate change and aesthetics (Lacy 2001:635–645). However, there does exist a wealth of scholarship of the various aspects of climate change and its interaction with pop-culture and movies or eco-cinema. As a corollary to the emergence of climate change fiction, there has emerged a number of monikers and neologisms associated with movies specializing in climate change like petro-fiction, solarpunk, Anthropocene fiction, solar fiction, climate trauma cinema, eco-trauma cinema, crisis cinema, Anthropocinema, envirotoon, etc. (Leikam and Leyda 2017; Svoboda 2014:43,111). For the purpose of this chapter, the use of the term eco-cinema embraces a wide range of transmedial cinematic interventions like documentaries, short films, movies, etc.

Eco-cinema has again been studied through various disciplinary prisms. There are contemporary Marxist perspectives which view pop culture including eco-cinema as a medium to demobilize and maintain docile populations. Similarly some view eco-cinema as a representation of the countercultural critique of technology-centric anthropocentrism, echoing the concerns of risk society and uncertainty theorists on the potential risks if not disasters of technology using the example of movies like Jurassic Park (von Mossner 2017; Lacy 2001:129,640). There are also rhetorical perspectives theorizing eco-apocalyptic cinema as an epideictic discourse offering a path to a hopeful future and redemption from our internalized guilt of climate change (Powell 2017:1). Alternatively feminism is another vibrant perspective which has brought to the fore the gendered assumptions of contemporary cli-fi cinema highlighting the heteronormative anxieties with respect to the nuclear family, gender roles and even the resilience of hegemonic masculinity even in supposedly dystopian feminist movies like *Mad Max, Fury Road*, etc. (Kendrick and Nagel 2020; Leyda *et al.* 2016:5,15). Other readings have linked the anxieties associated with the Anthropocene to the zombie genre movies (McReynolds 2015:149).

There are also substantial studies on eco-politics taking specific examples of movies as case studies including studies on the eco-politics of domicide in *Snowpiercer*, the impact on climatic discourse of *The Day after Tomorrow* (Leyda 2019:129). Similarly *Interstellar* has been critiqued for its politics of technological optimism and political defeatism, even upholding the frontier politics inherent in colonialism and even American exceptionalism couched in terms of mobility (Parson 2021: 82–112). The popular series *South Park*'s contested forms of engagement with climate change politics highlights the ideological battlegrounds drawing in pop culture reflecting the American society polarized between climate change engagement and denial (Wang 2013:23–24).

While most of these studies were concerned with the frame-based politics of representation of climate change, there is also substantial literature on the politics of reception. Audience research studies include the impact of climate change movies like *Day after Tomorrow*, *Age of Stupid*, etc., on various strata of the population including moviegoers, general public, etc. (Howell 2014; Sakellari 2015:70, 827). And, it is in this backdrop of the dominant eco-politics of most mainstream eco-cinema of the Global North, especially the U.S., that this chapter casts the spotlight on two documentaries — *Our Planet* and *Chasing Corals*.[1] The primary argument it makes is that they only reflect if not reinforce the existing dominant representations of climate change

---

[1]*Chasing Corals* can be found at https://www.youtube.com/watch?v=aGGBGcjdjXA. *Our Planet* is a documentary series which is available at Netflix and Youtube.

culminating in a largely ineffectual politics of transformative action by highlighting two prominent themes running through them.

Before proceeding with this analysis of eco-cinema, it is necessary to provide some conceptual clarification regarding analyzing documentaries against the backdrop of eco-cinema which mostly includes cli-fi as well. One could easily argue that documentaries have an explicitly didactic or educational focus, seeking to represent the "truth" as accurately as possible as opposed to movies. However, there are counterarguments that documentaries aren't different to movies in that they employ and manipulate images in accordance with the anthropocentric preoccupations of the directors. Nature documentaries construct nature in a certain political configuration. As Werner Herzog has argued "the boundary between 'fiction' and 'documentary' simply does not exist; they are all just films. Both take 'facts', characters, stories and play with them in the same kind of way" (Pegu 2017:55).

Chasing Corals illustrates the efforts of a crew of scientists, photographers, advertising professionals, divers, etc., to capture the beauty and the wide scale and dramatic climate change induced degeneration of coral reefs and their challenges in enhancing public awareness of this particular issue. *Our Planet*, on the other hand, is a documentary series that highlights the spectacular flora and fauna associated with a wide range of landscapes and the threats they face as a result of climate change. Here, this chapter identifies two prominent thematic threads running through which represents and serves to reinforce the dominant eco-political understandings of climate change prevalent in the Global North.

Temporality of Climate Change: Pre-trauma and Risk

The most dominant understandings of climate change exhibit a very distinct temporal form. Most of the dominant cinematic representations hinge on the politics of risk society or uncertainty theorists, whereby they experience severe anxieties of the future of climate change. In other words, they reflect a sense of "pre-trauma" anticipating a future dystopia. The temporality exhibited in these documentaries is no different. It largely follows a linear time-scale where there is the pristine or the "natural" followed by the unease of climate change culminating in glimpses of dystopia of destruction which are warned to get even worse. For instance consider the documentary *Chasing Corals*. The natural or the "past" is represented by extremely beautiful and colorful representations of nature. There are a number of shots highlighting the beauty of the corals and the myriad colors and hues of the flora associated with the fauna. There are themes of abundance and flourishing. *Our Planet* is no different highlighting the richness of the natural world illustrating the grandeur of the temporal rhythms of mating, preying, feeding, nesting, etc., of a wide variety of life-forms. This is then depicted to be followed by an intermediate phase of risk.

The reality is that climate change has ensured that life will never be the same in the natural world. There have always been severe environmental risks and inherent dangers which lurk beneath the "pristine" natural worlds. *Our Planet* highlights this with a number of "near-death" misses of the aesthetic subjects involved, including a seal-pup and a penguin which narrowly escaped being eaten by a polar bear and a whale. However *Chasing Corals* treats this aesthetics of unease in a rather cinematic manner. The corals before their inevitable perishing through bleaching resort to fluorescing resulting in what one of the crew refers to as "most vivid colours I had ever seen". It is even mentioned that this is an incredibly beautiful phase of death. However, even in this atmosphere of risk, "order still is maintained" through the natural climatic patterns which still continue. The climax or the expected dystopia is presented in glimpses. In *Chasing Corals*, it is represented by coral bleaching whereby the corals turn white. The degeneration is now complete. A crew member in the film laments, "It is dead as far as I can see; it is just algae and dead coral skeletons … flesh is rotting away". *Our Planet* sure depicts the catastrophe in dramatic ways. The mass deaths of walruses in Russia through their falls from great heights represent the aesthetics of destruction signifying the dystopia that the risk society is headed toward.

While such linear narratives present a powerful picture of climate change and has the potential to impact the viewer on an affective scale, its transformative political potential has been limited, at least so far. Given the unequal impact of climate change across the world, the linear temporal framework is largely the dominant view of those who are relatively less vulnerable to climate change. This is best highlighted in a later section on how climate change is not merely an anticipatory dystopia but an integral component of the everyday temporal existence for those who are the most vulnerable to climate change. This leads to the disruption of their temporal rhythms in unexpected manners.

Aesthetics of Excess: Charismatic Species, Sublime Landscapes, and Spectacles

Any discussion on the politics of climate change is incomplete if not impossible without aesthetics. Aesthetics, in the Rancieren sense, relates to the senses or experience that remain central to politics (Yusoff 2010:80). A significant challenge of the politics of action *vis-á-vis* climate change is making the very phenomenon of climate change visible given the multi-dimensional, imperceptible, invisible, and rather complex nature of climate change. This allows traction for influential sections of climate-skeptics and even climate-change deniers. On a philosophical level, this is also linked to the larger backdrop of the empirical legacy of Enlightenment where being seen matters.

A key orientation when it comes to eco-cinema, including documentaries and even cli-fi, is the didactic or educational approach (Leyda *et al.* 2016:6). Climate change cinema thus seeks to make the variegated, complicated, and the unseen aspects of climate change intelligible, accessible, and more importantly visible (von Mossner 2017:130). Thus, given different projections of the Great Acceleration of the Anthropocene which are revised on a regular basis, visual media, especially cli-fi, resort to a spectacular vision of a dystopia as the climax we are headed to. Unfortunately, some scenarios inevitably draw from the creative predispositions of the director, which aren't entirely scientifically accurate. Besides the simplified discourses of the climatic futures, this spectacular treatment has often been criticized as it feeds into the charge of climate alarmism often raised by climate skeptics (Svoboda 2014:56).

The point being made is that more often than not filmmakers resort to the politics of the aesthetics of excess. The aesthetics of excess, in simple terms, involves constant invocation of the "sublime". The "sublime" has a long and winding trajectory of meanings associated with it, ever since an imagined ancient era. However the overarching theme underlying the connotation of sublime can be termed as "defying description" or "beyond words". Thus the sublime is lofty and grand, invoking awe and veneration (Callahan 2020; Devetak 2005:10,622–629). The invocation of the sublime is often an intrinsic part of the visual eco-politics of climate change. It is not enough to make climate change visible, but it has to stand out. Another challenge is the right mode of making it stand out. The crew makers on witnessing the scale of coral bleaching are concerned because the sight of white corals everywhere actually looks beautiful and does not invoke the visual message of danger. The task is how to make such a beautiful appearance "appear" dangerous. The underlying assumption is that the sublime has to be invoked for meaningful public engagement with climate change.

As mentioned in the previous subsection, each phase of the temporal life cycle was accompanied by vivid manifestations of the sublime. Whether it was the flourishing of the corals, the pre-trauma signifying their death and finally their bleaching, it was accompanied by multiple hues and colors ('Overwhelmed in a reef'), the fluorescing ("most vivid colours … most beautiful transformation in nature") and finally the expanse of white, respectively. *Our Planet* takes the definition of "sublime" even a step further whereby various processes of temporal life rhythms of natural beings including mating, feeding, chasing, porpoising, and even dying are presented in a well-coordinated, rhythmic, and even choreographic manner. In addition one can find two other modalities of the sublime:

(a) Charismatic or mega-fauna and flora
(b) Pristine and "natural" landscapes

(a) *Charismatic mega fauna and flora*: The politics of conservation centered around certain flagship species, especially megafauna, has been well documented. While certain species, ranging from tigers and pandas to polar bears and penguins do have enormous symbolic capital of conservation owing to their aesthetically pleasing presence and charismatic appeal, they marginalize not only the conservation of other "lesser" aesthetically pleasing species and the larger biodiversity but also the people residing in such locales, especially indigenous communities (Jalais 2008; Taghioff and Menon 2010; Yusoff 2010:25,69,74–75).

In a sense, one could argue that the very intent of *Chasing Corals* was to raise the visibility of corals to a charismatic symbol of conservation. This was sought to be attained through lengthy shots of the beauty of corals in various hues and colors as well as the microscopic views of the coral, laced with colorful dyes. The beauty of various fauna and flora from clownfish to sea-dragons also enhance the charismatic appeal of corals. Similarly *Our Planet* fetishizes arguably the most charismatic species from each geographical zone. The frozen worlds are represented by king penguins, humpback whales, polar bears, albatrosses, and walruses. In the episode *Forests*, the Western Ghats is represented by merely the

Lion-tailed macaque and the hornbill, while the forests of Madagascar are represented by the idiosyncratic baobab tree and unique species like the fossa. In other words, charisma or aesthetic appeal emerges as arguably the most important prerequisite for representation in the politics of the "survival of the most charismatic".

The fetishization of charismatic mega species provides important insights into the subjectivities and anxieties of climate change in the Global North. What is projected here of utmost concern is the loss of "nature" which was once beautiful. *Chasing Corals* highlights the sense of loss and grief of "coral nerds" and "coral gurus" whose passions are linked with sea-diving and coral. It is the melancholy of the loss of beauty which hurts them the most. Jennifer Fay would refer to this as the illustration of "first-world middle class horror" of climate change (Baer 2018:82). This is not to say the destruction of corals wouldn't have ramifications. Coral reefs are natural barriers against sea erosion and cyclones and of course, are essential for the fisheries' ecosystem. The ramifications of their degradation would be severely impacting fishing communities and other vulnerable people across the world. However, the mention of such consequences barely take up around 10 seconds of the documentary spanning more than one and a half hours. What is chosen to be foregrounded is the middle-class aesthetic anxiety or horror of losing what was once beautiful and visually so appealing. In other words, reflecting much of the hegemonic politics of global politics, *Chasing Corals* casts a largely blind eye to the uneven impact of environmental degradation and focuses less on the questions of social justice and equity.

(b) *Critical eco-politics of "natural spaces"*: The role of eco-cinema in constituting and interpreting geopolitical spaces has been well documented (Carter and Dodds, 2014). One sees a similar mechanism at play explicitly so in the series *Our Planet*. Each episode is labeled according to the natural pristine spaces which have to be conserved like "frozen worlds", "jungles", "high seas", "forests", etc., thereby reinforcing and constituting the natural spaces which have to be conserved. The dynamic is a bit different in *Chasing Corals* whereby corals are constituted as the undersea equivalents of cities like Manhattan with constant comparisons of the intra-coral reef interactions to those occurring within cities. As mentioned before, they are presented in a pristine, sublime manner to invoke purity. There is a constant innovation of the splendor not only of landscapes through drone views but frequent satellite imagery to invoke the grandeur of Earth. The implication is that there is a "planet" or "nature" out there which is independent of human beings and needs to be conserved.

There have been a number of discursive studies highlighting how constant invocation of the splendor of a natural world only enhances the separation between man and the environment when both are intertwined. This separation reinforces the logic of relationship into a binary of either conservation or exploitation. Such forms of representation also have myriad implications. The focus is shifted overwhelmingly to conservation of pristine and beautiful landscapes thereby not merely marginalizing the more mundane landscapes but also marginalizing the concerns of the most marginalized and vulnerable and the issues of social justice. In addition, the sense of splendor reinforces the divide between us and the planet out there reinforcing our powerlessness and demobilizing transformative political action (Branston 2007:212). Lastly the constant tourist gaze could be detrimental for such landscapes as there could be a final rush to visit these landscapes before they perish forever (Branston 2007:219).

While melancholy about the future of charismatic species and expansive landscapes is an arguably effective mechanism and could result in changes on an affective plane, it does little to highlight the extractive technological excesses of capitalist modernity, which itself has been responsible for such conditions in the first place. For instance, *Chasing Corals* barely resorts to mere platitudes on raising children's awareness and clean energy and conservation. Besides the scant focus on the socio-historical and political undercurrents of climate change, these documentaries, similar to their cinematic counterparts, marginalize alternative eco-political imaginations, especially from the Global South. It is in this context, that one needs to take note of at least four major transmedial cinematic representations of climate change in India. This clearly features some of the alternative forms of eco-political imaginations, highlighting innovative and unconventional aesthetic modalities. The analysis below makes an attempt to elucidate these variations from the Global South.

Cinema of the Global South: Four Alternative Visions

The scholarship on media representations of climate change has been largely dominated by the Western cinema barring certain exceptions on audience reception and comparative perception analyses (Kakade 2013;

Matusitz and Payano 2011; Schäfer and Schlichting 2014:37,42,65). As an example, this section seeks to highlight four cinematic aesthetic visions of climate change from Indian cinema. This choice is based on a transmedial representation of media forms spanning documentaries, short films, and movies. The milieus and the impact of climate change are also distinct focusing on the varied impact of climate change on four diverse vulnerable locations — Himachal Pradesh, Odisha, Madhya Pradesh, and Kerala.

The Weeping Apple Tree[2] is a documentary that shows the plight of the apple farmers in Himachal Pradesh in the wake of climate change. *Climate's First Orphan*[3] is another documentary which highlights the lives of the coastal communities in Kendrapara, Odisha affected by the double burden of cyclones and climate change-induced coastal erosion. *Kadvi Hawa*[4] is a film which highlights the efforts of an old man to broker a deal with a bank employee, entrusted to recover loans in Bundelkhand, to save his son against the backdrop of the farmers' crisis and suicides exacerbated by climate change-induced drought. *Nelson*[5] meanwhile is a psychological thriller short film which examines the anxieties of the population of Kerala especially in the wake of the Kerala floods of 2018 and 2019 (amid widespread deforestation) which ravaged almost the entire state.

The selection of these cinematic forms represent a diverse range of media as well as varying manifestations of climate change impact at the ground level. The analysis that follow seeks to argue that these constitute representations of oppositional cinema from the Global South challenging the hegemonic forms of conformist eco-cinema from the Global North. Toward this endeavor, the following sections have identified two recurrent themes which run through these movies. These are (a) Adaptation and (b) Aesthetics.

(a) *Adaptation*: Most of the "conformist" cinematic representations of the West have largely focused on the question of mitigation of the negative impact of climate change. While adaptation remains a prominent theme in terms of strategy advocated within UNFCC, for some reason, it has been largely marginalized in visual media. All these movies selected deal with

---

[2] *The Weeping Apple Tree*: https://www.youtube.com/watch?v=Xc8yXYXZarY&t=497s.
[3] *Climate's First Orphan*: https://www.youtube.com/watch?v=TQH7AMuubTI.
[4] *Kadvi Hawa can be viewed on Zee5.*
[5] *Nelson*: https://www.youtube.com/watch?v=v9vsl3XFE8U&t=3s.

the question of adaptation in myriad ways. The optimistic language of adaptation inherent in most of the institutional frameworks obscures the pains often associated with this process of adaptation (Khan 2015:404).

Odisha remains particularly vulnerable to climate change which has exacerbated cyclones and its coastal erosion. The coastal communities world over, contributing very less to greenhouse gas emissions, are among the worst affected, suffering from domicide (even the loss of entire villages). In Odisha, owing to salination, they have suffered crop loss on an unprecedented scale resulting in extreme poverty. The adaptation here has been the most painful with most of the people depending on children who have migrated to urban areas and send them money to support their expenses. *Kadvi Hawa* presents a similar painful picture in central India where climate change induced droughts in an already infertile arid ravenous region that have driven many farmers to suicide. The adaptation here is in the form of temporary non-farm labor with barely subsistence wages. However adaptation isn't always a painful picture as the *Weeping Tree* illustrates. Among the apple farmers of Himachal Pradesh, the reality of adaptation has been more ambiguous. While climate change has been detrimental to the farmers in villages like Hurla (nestled in the lower slopes), it has been beneficial for apple farmers of villages like Nashala in the upper slopes of Himachal Pradesh. Besides highlighting the realities of adaptation, these cinematic forms show the complex and multidimensional picture of adaptation which is an uneven process generating a series of winners and losers, even in close proximity.

The process of adaptation to climate change is also very much contingent on the historical and socio-political realities of the spaces under consideration. These movies highlight the intersection of climate change with the historical and socio-political fabric of these regions. The climate change-induced droughts in Bundelkhand are exacerbated by the heartlessness of the impersonal bureaucratic institutions against the backdrop of bonded labor which has been prevalent for generations. The old man in the film thus repents: "Initially it was my father, now my son". Similarly in *Nelson*, climate change unfolds against the backdrop of the distrust of the government, the alleged timber industry-government nexus, lack of transparency, etc., which represent the socio-political fabric of Kerala. Lastly, whether it is the increasing alienation of the women in *Kadvi Hawa* amid the mundane house work which has only increased, or women who refuse to be married to the men of the affected villages of Kendrapara, climate change is indeed very much a gendered process.

(b) *Aesthetics*: As mentioned before, the documentaries from the Global North mirror the aesthetic treatment of much of cli-fi; invoking the constant aesthetic of excess. From the abundance, through risk and unease to the final destruction, there was a constant resort to the politics of the sublime using aesthetic landscapes as well as charismatic subjects. However, the subjects in the selected eco-cinema from the Global South engage with aesthetics in various interesting and unconventional registers. Consider *Nelson*, for instance that makes a representation of the anxieties of a state struggling to negotiate with its exceptionalism. Kerala as a state is founded on a sub-nationalism built around a discourse of exceptionalism interlinked with a faith in rationalistic modernity (Harikrishnan 2020; Mannathukkaren 2013:271). However, the impact of the climate change-induced floods was so drastic and sudden that it arguably changed the psyche of Kerala forever. It not only shook the exceptionalist pride but severely exposed the limits of the technological rationalistic modernity as deforestation and urbanization significantly worsened the costs of floods. *Nelson* is a reflection of this cultural anxiety. There is a severe distrust of the intentions of the government which is allegedly is hands-in-glove with the timber industry. The subjects also lament over the loss of agency and control, reflective of the helplessness towards the floods. One of the scientists entrusted with flood control laments the inefficiency of their measures "What we did so far was a myth".

In a sense, it is analogous to the Gothic response to the rationalistic legacy of Enlightenment. There is a constant dread and unease pervading the atmosphere with dogs turning hostile and trees producing mysterious substances. The body and the psyche of the protagonist Nelson is symbolic of the aesthetic terrors grappling (and continuing) the state. While the body has been sullied by the dirt, mud, and the floods, the temporal rhythms also have been disrupted. Far from the linear time-cycles of the subjects in the Western documentaries, *Nelson* presents a disjunctive temporal trajectory where the historical memory and the personal memory often intermix signifying confusion and chaos. The terror, confusion, and dread juxtaposed with the pristine visions of fauna in dreams and the sublime shots of forests present an alternate aesthetic challenge to the rationalist epistemologies signifying a Curtisian aesthetic of intuitive experiential knowledge.

In *Climate's First Orphans* the harrowing testimonies of domicile whereby everybody in the village points deep into the sea to locate their original homes ("My house was there") is juxtaposed with the mundane events of boys playing with the ball, girls chatting while filling water. Similarly in *Kadvi Hava*, there persists anxieties of those embroiled in the crisis ranging from the intra-familial relationships (father and son hardly talking anymore, the wife being alienated) to the innocent queries of the children on farmer deaths ("What happened to Janaki's father?") which are juxtaposed with the mundane of the everyday illustrated by people brushing teeth, carrying firewood, buying vegetables and children going to school. Personal memories associated with climate change again assume a high degree of significance, "Earlier there used to be two harvests a year, now there is nothing". There are unpleasant encounters with the body here as well. The old man, for instance, makes a comparison between people and rats as they either hang themselves or consume poison, highlighting the dehumanization in the most harrowing manner. Sure the unease descends into the dystopian climax of the suicide of the son (which the father sought to prevent at all costs), but then life continues as it does, with the guilt weighing heavily on the father. The treatment in these instances is far from a subliminal aesthetic; rather it signifies depiction of a post-disaster realism.

Conclusion

To conclude, this chapter examines the integration of the aesthetic, turned into climate change-based eco-cinema, taking the case-studies of two documentaries (*Chasing Corals* and *Our Planet*) which represents and reinforces the dominant understanding as also politics of climate change discourse that focusses more on charismatic species, conservation of beautiful landscapes in a specific linear temporality, infused with a sublime aesthetic of excess integral to each aspect. This was juxtaposed with four transmedial representations of climate change from India (as representative of Global South) taken to represent the impact of climate change through the travails of adaptation and a wide spectrum of aesthetic possibilities ranging from the mundane to the sublime, highlighting the impact on body, mind, and memory.

The post-pandemic era however has seen a wide range of climate change disasters ravaging both the Global North and Global South. They include, but are not limited to the heat domes in Canada, forest fires in Italy, Spain, Australian bush fires as well as unprecedented floods which ravaged Germany. The near universal impact of Coronavirus pandemic, which has resulted in economic recession across the major economies, has only enhanced the vulnerability of the Global North to climate change-induced disasters. What was once seen as anticipatory trauma has thus now become a regular reality of many locales even within the Global North. With the impact and vulnerabilities of climate change felt across the globe, the novel post-pandemic era has sounded a wake-up call for far more sensitive depiction and representation of climate change.

This is not to overstate the emancipatory potential of such representations from the Global South. The subliminal messaging of *Nelson* where there is a fear of the supernatural and notions of the revenge of nature obscures the role of the political and economic factors which resulted in the floods, thus depoliticizing the message. While the nuances of adaptation are highlighted in great depth, there is still a lack of focus on solutions. Further, there needs to be more audience-based studies or reception studies on whether alternative aesthetics results in transformative mobilization or aids climate change activism among the larger public. Finally, the dichotomy of the representations of Global North versus Global South cannot be generalized as there are exceptions in both scenarios. For example, while *Snowpiercer* was a harsh and direct critique of the very capitalist hierarchical modernity which had spawned the crisis, movies like *Princess Mononoke* from Japan showcased largely depoliticized mythical representations of climate change (Smith and Parsons 2012:25).

To sum up, the examples highlighted above illustrate various complicated facets of the process of adaptation which thus presents a far more nuanced account not only of the realities of climate change but of its uneven impact on the most vulnerable people bringing to the fore the aspects of social justice and inequity, which were hitherto marginalized in popular environmental discourses. Lastly, the unconventional aesthetically driven choices employed ranging from the painful memories and testimonies of the apple farmers of Himachal and the elderly people of Kendrapara and Bundelkhand to the disjointed memories and confusion highlight intuitive forms of experience-based knowledge which signify a reorientation toward a radically different ethics of climate change, forcing "new images full of loss and rage that scream through our aesthetic orders" (Yousuf 2015:93). A holistic understanding of the climate change, therefore, calls for moving beyond the dominant Euro-American representations of climate change to explore diverse representations from the Global South.

References

Åhäll, L. (2008). Images, popular culture, aesthetics, emotions: The future of international politics? *Political Perspectives, 3*(1), 1–44.

Baer, N. (2018). Cinema and the Anthropocene: A Conversation with Jennifer Fay, https://filmquarterly.org/2018/06/08/cinema-and-the-anthropocene/, accessed January 1, 2022.

Branston, G. (2007). The planet at the end of the world: "Event" cinema and the representability of climate change. *New Review of Film and Television Studies, 5*(2), 211–229.

Callahan, W. A. (2015). The visual turn in IR: Documentary filmmaking as a critical method. *Millennium: Journal of International Studies, 43*(3), 891–910. https://doi.org/10.1177/0305829815578767.

Callahan, W. A. (2020). *Sensible Politics: Visualizing International Relations* (1st edn.). New York: Oxford University Press.

Carter, S. and Dodds, K. (2014). *International Politics and Film: Space, Vision, Power*. New York: Columbia University Press.

Der Derian, J. (2010). Now we are all avatars. *Millennium: Journal of International Studies, 39*(1), 181–186.

Devetak, R. (2005). The gothic scene of international relations: Ghosts, monsters, terror and the sublime after September 11. *Review of International Studies, 31*(4), 621–643.

Dodds, K. and Hochscherf, T. (2020). The geopolitics of Nordic noir: Representations of current threats and vigilantes in contemporary Danish and Norwegian serial drama. *Nordicom Review, 41*(1), 43–61.

Kakade, O., Hiremath, S., and Raut, N. (2013). Role of media in creating awareness about climate change — A case study of Bijapur city. *IOSR Journal of Humanities and Social Science, 10*(1), 37–43.

Grayson, K., Davies, M., and Philpott, S. (2009). Pop goes IR? Researching the popular culture-world politics continuum. *Politics, 29*(3), 155–163.

Harikrishnan, S. (2020). Communicating Communism: Social spaces and the creation of a "progressive" public sphere in Kerala, India. *Open Access Journal for a Global Sustainable Information Society, 18*(1), 268–285.

Howell, R. A. (2014). Investigating the long-term impacts of climate change communications on individuals' attitudes and behavior. *Environment and Behavior, 46*(1), 70–101.

Jalais, A. (2008). Unmasking the cosmopolitan Tiger. *Nature and Culture, 3*(1), 25–40.

Hozic, A. A. (2014). Between "national" and "transnational": Film diffusion as world Politics. *International Studies Review, 16*(2), 229–239.

Khan, N. (2015). River and the corruption of memory. *Contributions to Indian Sociology, 49*(3), 389–409.

Kendrick, S. and Nagel, J. (2020). The cowboy scientist saves the planet: Hegemonic masculinity in cli-fi films. *Masculinities: A Journal of Identity and Culture, 14*, 5–34.

Lacy, M. J. (2001). Cinema and ecopolitics: Existence in the Jurassic Park. *Millennium: Journal of International Studies, 30*(3), 635–645.

Leikam, S. and Leyda, J. (2017). Cli-fi and American studies: An introduction. *Amerikastudien/American Studies, 62*(1), 109–114.

Leyda, J. (2019). Climate crisis, financial crisis: Negative mobility and domicide in 21st-Century American cinema. *Literary Geographies, 19*.

Leyda, J., Loock, K., Starre, A., Pinto Barbosa, T., and Rivera, M. (2016). The dystopian impulse of contemporary Cli-fi: Lessons and questions from a joint workshop of the IASS and the JFKI (FU Berlin). *IASS Working Paper.* https://doi.org/10.2312/IASS.2016.026.

Mannathukkaren, N. (2013). The rise of the national-popular and its limits: Communism and the cultural in Kerala. *Inter-Asia Cultural Studies, 14*(4), 494–518.

Matusitz, J. and Payano, P. (2011). The Bollywood in Indian and American perceptions: A comparative analysis. *India Quarterly: A Journal of International Affairs, 67*(1), 65–77.

McReynolds, P. (2015). Zombie cinema and the Anthropocene: Posthuman agency and embodiment at the end of the world. *Cinema 7: Journal of Philosophy and the Moving Image*, 149–168.

Parson, M. A. (2021). "Will God forgive us?": Christianity and the climate crisis in auteur cinema. Doctoral dissertation, Louisiana State University. https://digitalcommons.lsu.edu/cgi/viewcontent.cgi?article=6564&context=gradschool_dissertations.

Pegu, U. K. (2017). Cinema and the environment: An inquiry on ecocritical academia. *Desh Vikas, 4*(2), 51–60.

Powell, C. L. (2017). *Cli-Fi Cinema: An Epideictic Rhetoric of Blame* (Masters dissertation, Louisiana State University). https://digitalscholarship.unlv.edu/cgi/viewcontent.cgi?article=4029&context=thesesdissertations.

Rose, G. (2016). *Visual Methodologies: An Introduction to Researching with Visual Materials* (4th edn.). London: Sage.

Sakellari, M. (2015). Cinematic climate change, a promising perspective on climate change communication. *Public Understanding of Science, 24*(7), 827–841.

Schäfer, M. S. and Schlichting, I. (2014). Media representations of climate change: A meta-analysis of the research field. *Environmental Communication, 8*(2), 142–160.

Shapiro, M. J. (2009). *Cinematic Geopolitics*. New York: Routledge.

Smith, M. J. and Parsons, E. (2012). Animating child activism: Environmentalism and class politics in Ghibli's Princess Mononoke (1997) and Fox's Fern Gully (1992). *Continuum, 26*(1), 25–37.

Svoboda, M. (2014). A review of climate fiction (Cli-Fi) cinema … past and present. *Yale Climate Connections*, 1–20.

Taghioff, D. and Menon, A. (2010). Can a tiger change its stripes? The politics of conservation as translated in Mudumalai. *Economic and Political Weekly*, *45*(28), 69–76.

Wang, N. (2013). The currency of fantasy: Discourses of popular culture in international Relations. *International Studies: Interdisciplinary Political and Cultural Journal (IS)*, *15*(1), 21–33.

Weldes, J. and Rowley, C. (2020). So, how does popular culture relate to world politics? In Caso, F. and Hamilton, C. eds. *Popular Culture and World Politics. Theories, Methods, Pedagogies.* Bristol: E-International Relations.

Yusoff, K. (2010). Biopolitical economies and the political aesthetics of climate change. *Theory, Culture & Society*, *27*(2–3), 73–99.

von Mossner, A. W. (2017). Cli-fi and the feeling of risk. *Amerikastudien/ American Studies*, *62*(1), 129–138.

Chapter 4

Climate Change as a New Area of Sino-Quad Competition: Pacific Islands Perspectives

Artyom A. Garin*

Abstract

Given the combination of natural resources, unique geographical position, extensive Exclusive Economic Zones (EEZs), and the challenges facing the Pacific Island Countries (PIC), including economic and natural shocks, climate change is an urgent issue for the region, which have a negative impact on PICs' critical infrastructure as well as on food security.

However, there is another dimension of climate change — geostrategic. The intensification of the Sino-U.S. (incl. Quad) competition for leadership in the Indo-Pacific forces them to engage new areas of cooperation with small states in order to gain advantages in the region.

PICs are no exception. China, Australia, New Zealand, the United States, Japan, and Taiwan have been using Official Development Assistance (ODA) for decades to involve or keep PICs in their orbit.

* Research Assistant, Center of Southeast Asia, Australia and Oceania, Institute of Oriental Studies RAS, Russia, Moscow, ORCID ID: 0000-0003-4677-7221. E-mail: garinartyomalexeevich@gmail.com.

Australia, as the traditional leader of the South Pacific, has already secured its economic, defense, and vaccine presence in PICs. But the climate agenda, despite PICs' attempts to attract the attention of "senior" partners, could become a new platform for geostrategic competition. An important role in this issue is played by the commitment of U.S. President Joe Biden's administration to the climate agenda, which will affect the foreign policy vector of Australia in the same way that Donald Trump's foreign policy once affected the aggravation of Sino-Australian relations. However, Beijing can also offer green energy opportunities.

This chapter will provide a brief analysis of the climate change impact on the PICs economies, examine in detail the role of climate agenda in the foreign policy and ODA of the leading actors of the Indo-Pacific, as well as present the prospects for the rivalry of the PRC and Quad in the field of climate change in the South Pacific. This material contains an analysis of the PICs' behavior in the competitive conditions. Such processes have both practical and theoretical interest.

Keywords: PICs Climate Change, South Pacific

The Context: Australia and China's Influence in the South Pacific

The South Pacific is a vast territory of the globe. There are over 25,000 islands, 13 Pacific Island Countries (PICs), and over a dozen territories in associations with Australia, New Zealand, the U.S.A. and France. As of 2020, about 13 million people live in the PICs. This sub-region is traditionally called Oceania.

The South Pacific has a number of unique features. First, there is a tremendous influence of traditional actors — Australia, New Zealand, the United States, and France, — in the sub-region. Second, Small Island Developing States (SIDS) have extensive and potentially large reserves of natural resources (gold, copper, gas, oil, nickel, cobalt, wood) and are favorable positions for international trade.

Over the past decade, academics and politicians are increasingly focused on the growing influence of the PRC in the South Pacific. The precedence in the regional discourse is held by the defense as well as trade and economic agenda. However, the ecology is equally important for Oceania.

Many experts believe that close relations with China carry a potential threat to the sovereignty of SIDS. It is also widely believed that in this way China wants to expand its geostrategic space in the Indo-Pacific.

Despite the fact that Oceania is more remote from the potential theater of military operations in the Indo-Pacific than Southeast Asia or island states of the Indian Ocean, it still attracts more and more attention from leading regional and extra-regional actors. These include China, the U.S.A., Japan, France, India, and the U.K.

It is no secret that Australia's national interests are focused on Oceania. This sub-region has been a priority for Australia since the 19th century. The authorities of the Fifth Continent throughout history have wanted to gain control of the nearby island chains in order to avoid their capture by a potentially hostile power. In order to avoid internal instability in the region, Australia has developed the tools for providing Official Development Assistance (ODA) to the PICs. Subsequently, the effectiveness of aid was evaluated by other factors and joined the financing of Oceania. In recent years, China has taken leading positions in ODA and other areas.

It is notable that the PRC's official documents regularly mention South–South cooperation. In this way, Beijing wants to present itself as a developing state and a respectful partner for the countries of Oceania. Considering the trade dynamics, it is important to note that China's trade with PICs has grown 23 times since 2000 (from $270 million to $6.39 billion in 2018). In the period 2017–2018, China has overtaken Australia as the main trading partner for PICs (see Figure 1). The PRC was also the second donor of external aid in the sub-region. Such attention from China is due to several factors at once.

(1) Business interests, including access to resources (Papua New Guinea, Solomon Islands, etc.) and the Exclusive Economic Zone (EEZ), which allow more active participation in local fisheries (Kiribati, Solomon Islands, Samoa, Tonga, etc.);
(2) Important shipping routes that run along the northeast coast of Papua New Guinea, through the Solomon Islands and French Polynesia;
(3) The multiplicity of PICs that can provide support in international organizations.

Thus, the commercial interests helped China establish itself in the South Pacific. However, this was resisted by Australia and its partners, including Quad countries.

Bilateral trade between Australia/China and PICs (2000–2018, USD billion)

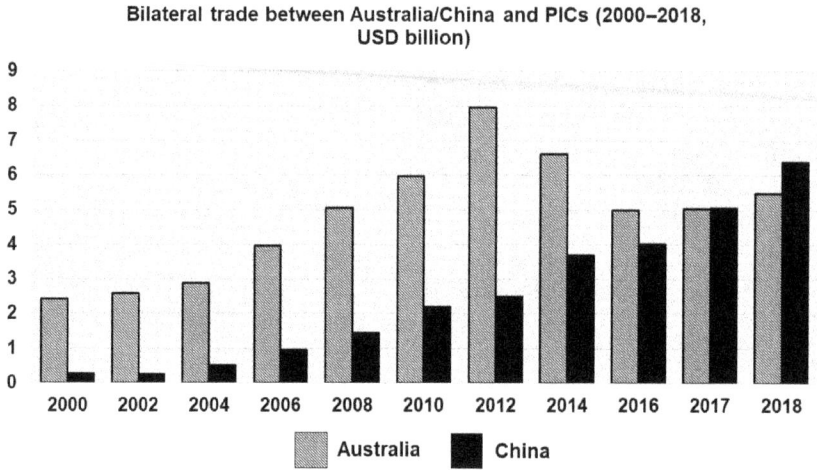

Figure 1:   Bilateral trade between Australia/China and PICs (2000–2018, USD billion).

Note: List of PICs: Papua New Guinea, Fiji, Solomon Islands, Vanuatu, Palau, Nauru, Kiribati, Federated States of Micronesia, French Polynesia, Cook Islands, New Caledonia, and Samoa.
Source: UN Comtrade (The UN 2021).

The rivalry between the two parties has already been reflected in PICs' official documents. In the Boe Declaration (Pacific Islands Forum 2018), they recognized "a dynamic geopolitical environment leading to an increasingly crowded and complex region". In addition, PICs agreed that national security plays an important role in the Oceania.

Among the provisions of the declaration, we can also find a clause on "respect and assert the sovereign right of every Member to conduct its national affairs free of external interference and coercion" (*Ibid.*). Given the rhetoric of Canberra, as the leading actor in the sub-region, such a statement can be perceived as a message to China. However, this could also be a sign to Australia, the U.S. and, to a lesser degree, New Zealand.

The Meaning of "Security" for PICs and Its Foreign Policy Strategies

Before analyzing the impact of climate change on PICs, it is necessary to consider how these countries perceive the rivalry between Quad and the PRC. In particular, the strategies of Oceania countries in this situation deserve special attention.

To begin with, the term "security" has a slightly different character for the PICs. Within the framework of multilateral meetings, issues of governance, healthcare, education, infrastructure, etc., are more often discussed. Natural disasters caused by climate change are considered the main threat in the sub-region. Traditional actors of the South Pacific are actually not able to fully support such a vision of PICs.

However, they are sensitive to the capabilities of larger actors. This also applies to foreign policy. PICs understand that they are inside a competitive environment, where large actors can be expected not only to try to develop ties but also to exert pressure. It is no secret that their "senior partners" perceive SIDS as weak actors. Over the past decades, Australia, the United States, and New Zealand decided that the sub-regional order depends on themselves.

Due to the transformation of the regional order in the Indo-Pacific, the PICs' geostrategic landscape is becoming more complex and multidimensional. They are forced to adapt foreign policy strategies to the new conditions in which the region is currently home to 4 of the world's 10 largest military spending countries. Asia and Oceania increased its global share of military spending from 19.2% in 2010 to 27.7% in 2019 (Swedish Defence Research Agency 2021).

Australia is also taking action. According to official data, in 2019–2020, the government allocated $38.7 billion for defense (Department of Defence 2019), and in mid-2020 provided another $397.4 billion for the next 10 years for the modernization of the armed forces and the military-industrial complex (Janes 2020). As Australian Prime Minister Scott Morrison said, the country should be better prepared for the prospect of a "high-intensity conflict" (Prime Minister of Australia 2020).

The United States, in turn, proposed to establish a new base in the Pacific (*The World* 2021). Washington and Canberra are working together on the Manus Wharf project in the northern part of Papua New Guinea.

Gradually, the PRC is increasing its power in the sub-region. Beijing has expressed its readiness to send anti-riot equipment and an *ad hoc* police advisory team to the Solomon Islands to help diffuse the civil unrest (*South China Morning Post* 2021). In addition, China has offered New Zealand to develop defense cooperation (Ministry of National Defense 2022).

This situation in the Indo-Pacific became a strategic shock for small actors. The rivalry of the great and middle powers for leading positions in the region requires a response from PICs. In addition, the population of

Oceania is aware of the consequences of such a serious conflict as the World War II. Due to such historical experience PICs' responses are more peaceful.

The PICs' leaders themselves understand the need to diversify international relations. This is probably why the PICs decided to take a more balanced position, so as not to be part of the states potentially participating in the rivalry with Beijing. For example, PNG Prime Minister James Marape refuted statements in the media allegedly made by the commander of the PNG Defense Forces, Major General Gilbert Toropo, that the growing presence of China is a challenge to PNG security. It is not for nothing that Oceania has a well-established foreign policy vision: "Friends to all and enemies to none". Most PICs share similar national interests, which is why they are considered as an integral whole.

Nowadays, PICs use hedging or isolation strategies. They also have the opportunity to build a balance of power. It could be expressed in two dimensions. In the first, any of the PIC's can join the Quad Plus or the PRC side. Or PICs can unite and promote their vision of a sub-regional space without the larger actors. Thus, they will be able to ensure that the sub-regional environment is competitive for all countries, and not only for great or middle powers.

However, both scenarios have low probability. During the pandemic, PICs will not take side of any actor, while maintaining the ability to receive assistance from various sources; in particular, to fight against the consequences of climate change.

Climate Change's Impact on PICs

The PICs are highly susceptible to natural disasters that negatively affect their welfare. The International Monetary Fund (IMF) has published a study on the economic consequences of natural disasters in Oceania. According to scientists, a natural disaster in the short term can decrease GDP growth by 2%, and also increase the trade deficit and government spending by 5% of GDP (Lee, Zhang, and Nguyen 2018).

It is widely believed that the states of Oceania are homogenous, but they have many differences. For example, the soils. They are not only imbued with volcanic soil, diverse terrain, and comfortable altitude above sea level (for example, Papua New Guinea, Fiji, Solomon Islands, Vanuatu, etc.) but also low-lying coral atolls (for example, Kiribati,

Marshall Islands, Cook Islands) in the South Pacific. The first group of countries is mainly located in Melanesia, has vast land territories and fertile lands, while most of the atolls are spread over the territory of Micronesia and Polynesia, have a small land area and are often unsuitable for agriculture. However, in conditions of a lack of fertile land, the local population is trying to benefit from the marine EEZ. Fish is almost the only source of animal protein in Tokelau, Kiribati, and Palau.

The PIC's population also correlates with the land area. The number of inhabitants of PNG is about 9 million people. It exceeds the population of all other PICs. According to the United Nations (UN), the total population of Oceania will increase by 56% in 2050 (The UN 2021). It poses new challenges to the sub-region, especially to Papua New Guinea. The countries need to produce enough food to feed a growing number of citizens and ensure an adequate level of food security.

Rural residents reside in large numbers in the Oceania. In particular, in Papua New Guinea, the Solomon Islands, and Vanuatu, almost 80% of citizens live in rural areas and rely on properly grown crops. As a result, agriculture makes a significant contribution (up to 28% of GDP) to the economies of PICs, along with resource extraction, tourism, or remittances from seasonal workers.

Forests play a special role in Oceania. Timber is an important export product of Melanesian countries. In particular, this is true for Papua New Guinea and the Solomon Islands. In 2019, rough wood accounted for 5.78% of PNG exports for $638 million (OEC 2021). Wood also consists 65.8% of Solomon Islands' exports and exports were worth $419 million (OEC 2021). This sub-region is home to rare species of trees; these are prohibited for export by foreign companies. However, despite the prohibitions, forests are being illegally razed. As a consequence, the unique nature of Oceania suffers. PNG currently hosts 30% of the balance of tropical rainforest in the world (S&P Global Platts 2021). Local politicians are struggling with this problem. A major regional actor that will support them will be able to earn the trust of PICs leaders.

The waters of Oceania contain some of the largest tuna resources in the world. However, given the combination of vast EEZ and rather limited material resources, it is feasible to ensure proper supervision of the natural resources only on a multilateral basis. Sea level also changes regularly. Accordingly, the territory of PICs is also changing. This creates new challenges at the level of international law. In 2010, the PIF developed the Pacific Ocean Landscape Concept, which outlines the need for "an ongoing regional effort to fix baselines and maritime boundaries". It is made "to ensure the impact of climate change and sea-level rise does not result in reduced jurisdiction".

Moreover, PICs' vulnerability to natural disasters reduces their ability to repay large loans. For example, Tropical Cyclone Gita hit Tonga in 2018 and affected 80% of the Polynesian country's population (The UN 2018). Thus, climate change can become an obstacle to the sustainable development of PICs.

The "Green" Rivalry between Quad and China in the South Pacific

Hurricanes and storms have a disastrous impact on infrastructure, food security and level of living. At the same time, most PICs are faced with the need for transport connectivity of the territories, which can ensure the access of farm products to larger domestic and foreign markets. Under the circumstances, external financing is crucial for the ecology of the sub-region. Over the past 10 years, i.e., 2010–2020, the amount of aid provided to Oceania countries was $26.47 billion. ODA related to agriculture, fisheries, and forestry, as well as water resources and the humanitarian sphere, accounts for 15.4% of this amount. A large share was meant for education, healthcare, civil society, and transport infrastructure. Here it is important to share an example of the struggle of PICs' leading donors for trust in the area of natural disasters and climate change. The comparison of the Quad and China is especially relevant in the aggravation of their relations and the growing economic presence of China in the South Pacific.

Australia allocates large funds to prevent the consequences of natural disasters. In March 2015, Cyclone Pam hit Vanuatu. The disaster affected almost 2 million people and damaged the infrastructure, including schools and hospitals. As a result, Australia allocated about $50 million dollars and held preliminary consultations with the Government of Vanuatu. The assistance contributed to the construction of 71 medical institutions, 95 classrooms, 51 public buildings, 80 water supply systems, 27 tourist bungalows, and 6 livestock centers. In 2017 Australia has granted $5.5 million for the evacuation and resettlement of more than 10,000 people from the eruption of a volcano on Aoba Island (Department of Foreign Affairs and Trade 2017).

China has also supported these island nations. As stated in its new white paper on foreign aid: "China emphasizes the importance of comprehensive recovery schemes for disaster affected countries, providing systematic reconstruction support". In addition, the PRC implements targeted but large-scale food and economic projects in Oceania. The Pacific Marine Industrial Zone (PMIZ) was China's first attempt to create an SEZ in Oceania, as well as the first SEZ in the history of Papua New Guinea. However, the PMIZ initiative did not cause as much resonance as the memorandum of the PRC and PNG on the construction of a multifunctional fishing industrial park worth about $150–$200 million on Daru Island.

It needs to be stressed here that the project is located on the southern coast of Papua New Guinea at the Torres Strait, literally 200 km from Australia. Against the background of the aggravation of Australia–China relations, the Chinese infrastructure initiative has raised concerns among a number of Australian experts. First, they feared extensive fishing by Chinese fishing vessels. Second, according to Canberra, the construction of a Chinese port could become a threat to Australia's national security. It is believed that in case of conflict, the Chinese industrial park could be converted into a naval base. China's Ambassador to Papua New Guinea Xue Bing rejected apprehensions that there are ulterior motives on the Chinese side and said that such investments "will definitely enhance PNG's ability to comprehensively develop and utilise its own fishery resources" (Radio New Zealand 2021). At the same time, he stressed that commercial cooperation between two sovereign states (PNG and China) does not require the prior consent of a third party (*Ibid.*).

It did not stop with Papua New Guinea. In September 2020, information appeared about China's possible land reclamation works in Kiribati. It is the island state in the heart of Polynesia, which is simultaneously located in four hemispheres of the Earth and has rich fish reserves and a huge 3.5 million square kilometer EEZ. In September 2019, China and Kiribati resumed their diplomatic relations, and Taiwan announced the severance of ties with the Polynesian state. Kiribati has for long attracted the international community to climate change impacts, as its atolls are gradually absorbed by the Pacific waters.

The period 2020–2021 has become the starting point for the recognition of the geostrategic significance of climate change. The leading Indo-Pacific actors have come to a partial understanding that the Cold War military tools are not effective in the recent times. It is quite difficult to gain full trust by supplying weapons, while assistance in response to real global challenges will be more productive.

Moreover, the PRC effectively reacted to this aspect. Contrary to popular belief that China's assistance is limited, Beijing's understanding of PICs' pressing problems has improved over the years. In the official statements, China declares its support for PICs' sustainable development efforts. In Oceania, there is a requirement for a more independent foreign policy. In this way, China emphasizes respect to the sovereignty and territorial integrity, as well as the "independent choice of development paths" of PICs (Embassy of the People's Republic of China in The Republic of Fiji 2021).

The creation of AUKUS between Australia, the United States, and the United Kingdom allowed the PRC to use the maritime security and the nuclear issue agenda. During the China-Pacific Island Countries Foreign Ministers' Meeting, Wang Yi highlighted the "vital importance to the people's health and safety as well as sustainable economic development of the Pacific Island Countries" (*Ibid.*). This concerns solving "the issue of discharging Advanced Liquid Processing System (ALPS) treated water into the ocean" and "the South Pacific Nuclear Free Zone" (*Ibid.*).

The Oceania states have also expressed readiness to support and join the Global Development Initiative proposed by President Xi Jinping at the general debate of the 76th session of the UN General Assembly. It is expected that this initiative will "ensure alignment with the Pacific Roadmap for sustainable development and the 2050 Strategy for the Blue Pacific Continent".

In addition, China has also announced that it will "set up a China-Pacific Island Countries climate action cooperation center" and continue to assist PICs. Moreover, it will be carried out under the framework of South–South cooperation. Given the reformatting of China's aid policy, it should be expected an influx of funding to climate change.

Australia is also trying to keep pace. The Indo-Pacific Carbon Offsets Scheme was initiated for this purpose. It includes assistance in delivering "renewable energy and nature-based solutions projects" as well as $104 million investments over the next 10 years. The two most populous countries of Oceania — Papua New Guinea and Fiji — have already joined the scheme. For example, PNG Minister for environment, conservation, and climate change Wera Mori gave some indication of future plans at growing trees and mangroves in the country (S&P Global Platts 2021).

Canberra is also counted on for assistance from its partners, who are also trying to contain China's influence growth in the Indo-Pacific. The Infrastructure for Resilient Island States was announced at the 26th UN Climate Change Conference in Glasgow. The IRIS includes India, Australia, and the United Kingdom. The key areas of activity of the IRIS parties are highlighted here.

India will continue its focus on the Indian Ocean Island Countries. Among the PICs, the most suitable countries are Fiji and Papua New Guinea in Melanesia. Given the fact that about 40% of the population of Fiji is of Indian origin, New Delhi has a unique opportunity to interact with the Indian diaspora in Fiji and thereby achieve economic and political results. The support of the diaspora in Fiji can have a positive impact in the medium and long term on ensuring stronger India's influence in the South Pacific.

In the case of Australia, it will be mainly the South Pacific, although Canberra also has mechanisms for cooperation in the Indian Ocean with India. Around 15 years ago, the authorities of the Fifth Continent allocated about $60 million to provide humanitarian aid to Sri Lanka, Maldives, Seychelles, etc. During the period 2020–2024, Australia plans to provide $1.4 million ODA to the Indian Ocean Island Countries.

The Indo-Pacific as a dual geostrategic space also contributed to the involvement of the U.K. Because of Brexit, London is still trying to develop its own network of ties all over the world. The U.K. wants to restore its position in wider Indo-Pacific, in particular with the "Pacific Uplift" strategy. It has already opened three new High Representative Offices in Samoa, Tonga, and Vanuatu. Britain has an overseas territory in Oceania — Pitcairn Islands. COP 26 in Glasgow, which was attended by many PICs' representatives, also benefited the U.K.

India, Australia, and the U.K. are paying attention to island countries for several reasons. First, there are 58 SIDS in the world. Their support can provide large-scale support to any initiative. Second, Australia, India, the U.K., and other Quad partners try to prevent China's rise in their traditional zones of influence.

IRIS can develop in at least two directions. Either India, Australia, and the U.K. will provide ODA to small island developing states together, or each will act in a pre-agreed sub-region without interfering with each other's activities. In any case, joining IRIS is the important trust factor among Indo-Pacific donors.

Potentially, support for IRIS can be ensured by Japan. Over the last 10 years, it has provided $1.8 billion ODA to PICs. In 2014, Japan upgraded the Port of Betio in Kiribati. The project cost amounted to $36.94 million. However, while it was engaged in the project, the Australian media did not concern itself about Japan's plans to secure a naval presence in the sub-region, using new ports to realize its military ambitions. With similar projects in China, the Quad's reaction was more critical.

Japan is often followed by Taiwan. In the late 20th and early 21st centuries, China has faced diplomatic competition from Taiwan. As a result, Oceania has become one of the key areas of their confrontation. To involve PICs in their orbit, the PRC and Taiwan took various initiatives and also used the funding to achieve foreign policy goals. The active phase of the rivalry between China and Taiwan in the South Pacific lasted until 2008, when the global financial crisis occurred.

It is now that history is repeating itself, but in a different format. Major Indo-Pacific donors are competing for recognition as a power that will dominate the future post-COVID world.

The U.S. connection to IRIS or other Quad-countries' climate projects could provoke debates about the growing competition with China. However, in the mid-term, Washington will enhance cooperation with SIDS. The position of Director for Oceania has already been created in the U.S. National Security Council. In addition, the U.S. still needs to develop a strategy for cooperation with the countries of Micronesia after 2030, when a number of agreements between the parties expire.

IRIS may also complement the Supply Chain Resilience Initiative (SCRI). In April 2021, the trade ministers of Australia, India, and Japan announced its official launch. According to official documents, the establishment of SCRI was a response to supply chain disruptions caused by the consequences of the COVID-19 pandemic. However, based on the composition of the grouping, it can be assumed that SCRI is a project of alternative supply chains designed to reduce economic dependence on China. The SCRI parties still don't have a clear strategy for implementing the stated tasks and there are a number of uncertain formulations.

Each IRIS country is ready to allocate about $10 million for the projects. Despite the fact, that this is a small sum by modern standards, the SIDS will be pleased to receive aid. Australia, Japan, the United States, and a number of other states are interested in deterring China's climate ambitions in SIDS. Nevertheless, as soon as talks begin about "climate coalition" for competition with Beijing, it will become more difficult to involve small actors in the initiatives. Moving forward, multilateral cooperation, especially through the United Nations, will be more effective.

Conclusion

The features of PICs have affirmed the need for a wider approach to security, because climate change is the main threat to the South Pacific. Small actors are struggling with the crisis in various areas, so it is possible to win their hearts and minds only by creating, and not by escalating the situation in the sub-region.

The recognition of climate change as a national security imperative suggests that environmental issues will assume a prominent place in the perspectives of PICs. In turn, increased competition in the Indo-Pacific has made ecology and natural disasters as areas of close attention for larger Indo-Pacific actors. A special role here is also played by the factor of China, which has already demonstrated its powerful opportunities to promote the environmental agenda.

Despite discussions about potential PLA naval bases in the South Pacific, these have not appeared there. The United States and Australia undertook the reconstruction of Manus naval base to consolidate their influence in the sub-region. According to the final communique of the AUSMIN meeting in September 2021, Australia is ready to act as a fulcrum for the United States if it is necessary to project forces against China in Asia. Surely this could provoke further militarization of the South Pacific, but the likelihood of an interstate conflict in the region is too small. The local population and leaders are simply not interested in further confrontation.

Despite Australia's special attitude to CO_2 emissions, it is forced to pay more attention to climate change. At least two dimensions have had an impact on the change of Canberra's policy direction: foreign policy and humanitarian considerations.

First, due to the commitment of the current government of Scott Morrison to the U.S. foreign policy, Australia has no choice but to pay more attention to the green agenda, which is close to the Joe Biden administration and is an essential component of the Quadrilateral Security Dialogue.

Second, Australia has traditionally paid little attention to the climate agenda of PICs, which are actually Canberra's zone of influence. As the trends of climate change and sea level rise intensify, climate aid may take a stronger position in Australia's relations with the SIDS. If it hesitates to do so, China will take the leading role in this issue. The PRC is looking for new points of contact with Oceania. China is establishing cooperation centers on climate change, poverty reduction, and economic development of PICs. Without Beijing's resources, it will be very difficult to resist climate change in the sub-region and the world.

Quad's rivalry with Beijing in the humanitarian and climate agenda will intensify. Given the entire range of countries that have united to compete with Beijing, it may also attract new states to the region. All this could help the SIDS diversify their relations, but will make the choice of the executors of climate projects more difficult. Participation in any initiative may be perceived as a choice by China or Quad. On the other hand, with PICs' assistance, one of the sides will be able to provide itself with greater strategic depth.

Acknowledgments

The author thanks the Association of Asia Scholars (AAS) and Saurabh Thakur for expert support for this chapter. I'm also grateful to Swaran Singh (Chairman and Professor in the Centre for International Politics, Organization and Disarmament (CIPOD), School of International Studies of Jawaharlal Nehru University, President of Association of Asia Scholars) and Reena Marwah (Professor in the Department of Commerce, Jesus and Mary College, Delhi University) for their valuable assistance during the preparing of this chapter.

References

A safer Australia — Budget 2019–20 — Defence overview (2019). Australian Government, Department of Defence, accessed on November 10, 2021. https://www.minister.defence.gov.au/minister/cpyne/media-releases/safer-australia-budget-2019-20-defence-overview.

Australian Government, Department of Foreign Affairs and Trade (2018). Development assistance in Vanuatu, Supporting cyclone recovery and reconstruction in Vanuatu, accessed on November 10, 2021. https://www.dfat.gov.au/geo/vanuatu/development-assistance/Pages/supporting-cyclone-recovery-reconstruction-vanuatu.

Boe Declaration on Regional Security (2018). Pacific Islands Forum, accessed on November 10, 2021. https://www.forumsec.org/2018/09/05/boe-declaration-on-regional-security/.

China's International Development Cooperation in the New Era (2021). The State Council Information Office of the People's Republic of China, accessed on November 10, 2021. http://www.scio.gov.cn/zfbps/32832/Document/1696686/1696686.htm.

Embassy of the People's Republic of China in The Republic of Fiji (2021). Joint Statement of China-Pacific Island Countries Foreign Ministers' Meeting, accessed on November 10. https://www.mfa.gov.cn/ce/cefj//eng/sgxw/t1916195.htm.

Growing at a slower pace, world population is expected to reach 9.7 billion in 2050 and could peak at nearly 11 billion around 2100. (2021). The United Nations, Department of Economic and Social Affairs, accessed on November 10, 2021, https://www.un.org/development/desa/en/news/population/world-population-prospects-2019.html.

Janes (2020). Australia details long-term defence funding plans, accessed on November 10, 2021. https://www.janes.com/defence-news/news-detail/australia-details-long-term-defence-funding-plans.

Lee, D., Zhang, H., and Nguyen, C. (2018). The economic impact of natural disasters in Pacific Island Countries: Adaptation and preparedness. IMF Workpapers, accessed on November 10, 2021. https://www.imf.org/en/Publications/WP/Issues/2018/05/10/The-Economic-Impact-of-Natural-Disasters-in-Pacific-Island-Countries-Adaptation-and-45826.

Ministry of National Defense of the People's Republic of China (2022). China to promote military relations with New Zealand: Defense Spokesperson, accessed on January 28, 2022. http://eng.mod.gov.cn/news/2022-01/27/content_4903786.htm.

Ocean Accounting for Disaster Resilience in the Pacific SIDS (2018). The United Nations, Economic and Social Commission for Asia and the Pacific, accessed on November 10, 2021. https://www.unescap.org/sites/default/d8files/knowledge-products/Ocean%20accounts_30Oct2018_LowRes.pdf.

OEC (2021). The observatory of economic complexity, accessed on November 10, 2021. https://oec.world/en/profile/country/png.

Prime Minister of Australia (2020). Defending Australia and its interests, accessed on November 10, 2021. https://www.pm.gov.au/media/defending-australia-and-its-interests.

Prime Minister Marape dispels perception of China as a Security Threat to PNG (2021). Department of Prime Minister and National Executive Council, accessed on November 10, 2021. https://www.pmnec.gov.pg/index.php/secretariats/pm-media-statements/319-prime-minister-marape-dispels-perception-of-china-as-a-security-threat-to-png.

Radio New Zealand (2021). Chinese cast net over neglected PNG border zone, accessed on November 10, 2021. https://www.rnz.co.nz/international/pacific-news/433180/chinese-cast-net-over-neglected-png-border-zone.

Regional Defence Economic Outlook 2021 Asia and Oceania (2021). Swedish Defence Research Agency, accessed on November 10, 2021. https://www.foi.se/rest-api/report/FOI%20Memo%207532.

S&P Global Platts (2021). COP26: Papua New Guinea joins Indo-Pacific carbon offsets scheme initiated by Australia, accessed on November 10, 2021. https://www.spglobal.com/platts/ru/market-insights/latest-news/energy-transition/110521-cop26-papua-new-guinea-joins-indo-pacific-carbon-off-sets-scheme-initiated-by-australia.

South China Morning Post (2021). Beijing has expressed its readiness to send anti-riot equipment and an ad hoc police advisory team to the Solomon Islands to help defuse the civil unrest, accessed on January 1, 2022. https://www.scmp.com/news/china/diplomacy/article/3160946/china-sends-anti-riot-gear-and-police-advisers-solomon-islands.

The World (2021). The US is building a military base in the middle of the Pacific Ocean. Micronesian residents have questions, accessed on November 10, 2021. https://theworld.org/stories/2021-08-24/us-building-military-base-middle-pacific-ocean-micronesian-residents-have.

UN Comtrade Database (2021). The United Nations, accessed on November 10, 2021. https://comtrade.un.org/data/.

https://doi.org/10.1142/9789811263750_0005

Chapter 5

Political Economy of River Ecocide in Bangladesh: A Study in the Context of Dhaleshwari River

Rabby Us Suny*, Oliver Tirtho Sarkar*, and Md Abid Hasan*

Abstract

This chapter scrutinizes the political-economic factors responsible for the river ecocide that augments the impact of climate change. Ecocide, the deliberate or thoughtless destruction of the environment, has existed throughout the human history. However, it has never been as imminent as it has been in contemporary times. Activities ranging from industrial manufacturing to daily household chores are alarmingly contributing to the phenomenon. Most of the prevailing studies focus on the impact on the environment by industrial activities but do not emphasize the issue of ecocide. Therefore, this chapter will offer a diverse spectrum in addressing river ecocide in Bangladesh through the lens of political economy. Firstly, the research evaluates the current political-economic influences that enhance the practice of ecocide. This chapter seeks to explore the impact of ecocide on people's livelihood and socio-environmental sustainability.

---

*Rabby Us Suny is currently working as an 'Annotation Specialist' in Augmedix Bangladesh Limited. Oliver Tirtho Sarkar is a Lecturer in the Department of Environment and Development Studies at United International University. Md Abid Hasan is a development professional and independent researcher.

Furthermore, the study probes into the long-term effects of the ecocide that might increase the impacts of climate change. Findings of the study show that lack of efficient implementation mechanism allows the tannery industry to cause river ecocide although environmental laws and policies are in place. To conclude, the study recommends a responsible, inclusive, and collaborative approach that will tackle the issue of ecocide and prepare Bangladesh to fight climate change constructively.

Keywords: Climate Change, Leather Industry, Power Dynamics, Sustainability, Tannery

Introduction

Ecocide, one of the well-known buzzwords of the time, has been coined as a term during the Vietnam War (in the 1970s) by a group of scientists led by Arthur Galston, an American biologist (Zierler 2011:2). Ecocide simply means intentional destruction of the (natural) environment (David 2020:1). Schwegler (2017:73) has also defined ecocide as crimes against peace that involve the right to life of the earth's inhabitants, including human beings, flora, and fauna. According to the Stop Ecocide International, ocean damage, deforestation, land and water contamination, air pollution, etc., are prominent examples of ecocide. To cite a few of the most prominent incidents of ecocide in history, Agent Orange (1961–1971, Vietnam; the first-ever incident marked as ecocide) (Chang *et al.* 2017:633), the disappearance of the Aral Sea (from the 1980s to date, Central Asia) (Loodin 2020:2495), oil exploitation in the Niger delta (a mangrove ecosystem) (ongoing for almost 100 years in Nigeria) (Onyena and Sam 2020:3), Palm oil production in Indonesia (from the early 70s to date) (Maskun *et al.* 2021:3), etc., can be mentioned.

Recently, in 2019, the devastating wildfire in the Amazon Rainforest drew global attention. The Bolsonaro administration of Brazil patronized extractive economic development that included agribusinesses, hydro-electric dams, mineral entrepreneurship, and even the displacement of indigenous culture in the rainforest. Resultantly, the international corporations criticized the Brazilian government for attempts to colonize the indigenous culture by unleashing economic development (Raftopoulos and Morley 2020:1617–1618). Another report showed the power dynamics of palm oil companies and their ecocide on La Pasión River, Guatemala, where the contamination by those companies caused the death of thousands of fish and forced the local community to cope with the situation (Zepeda 2017:3). The report also indicated that the victims were not given sufficient support; instead, their condition worsened by the food scarcity and health hazards.

In this era of sustainable development, understanding ecocide (and behaving accordingly) is crucial. Among the sustainable development goals (SDGs), goal 12 (responsible consumption and production), goal 13 (climate action), goal 14 (life below water), and goal 15 (life on land) are highly interlinked with the concept of ecocide. According to the Sustainable Development Goals Report 2020, the performance of Bangladesh is on track for goal 13, stagnating for goal 14, decreasing for goal 15, and no data was available for goal 12 (Sachs *et al.* 2020:40). If the notion of ecocide can be effectively considered, that would undoubtedly help Bangladesh achieve its sustainability goals.

However, climate change cannot be ignored as well since Bangladesh is one of the worst victims (Naser *et al.* 2019:1–2). According to a 2018 study, Bangladesh is facing difficulties coping with the ongoing climate changes (Ministry of Foreign Affairs of the Netherlands 2018:3). The impact of climate change is visible through the rapid changes, such as high temperature, unusual rainfall, frequent cyclones, sudden flooding, and droughts each year that make Bangladesh significantly vulnerable (Luetz 2018:61). Almost two decades ago, Agrawala *et al.* (2003:6) had predicted that Bangladesh would likely experience a 1.40°C temperature rise by 2050 and a 2.40°C rise by 2100 that indicated a disastrous future for Bangladesh. The nature and environment of the country have been its bio-shield for a very long time (Rahman 2018:314–316). If Bangladesh fails to protect the environment, it will lead to climatic hazards.

The concept of ecocide is not prevalent in Bangladesh; however, ecocide exists without the label. During the previous literature search, few reported cases have been found. Establishing a safari park in the Lathitila forest, Moulvibazar, a dying forest in the Indo-Burma Biodiversity Hotspot, is being considered ecocide (Akash 2021:July 7). The construction of the Kaptai Dam, the only hydroelectric power station in the country, is one of the oldest and cruelest examples of ecocide in Bangladesh (Sayeda 2021:April 17). Among the recent incidents, the proposed Rampal coal-based power plant near the Sundarbans is a perfect example of (to be) ecocide (Sayeda 2021:April 17).

Since the concept of ecocide was not widespread in Bangladesh until recently, previous studies about river ecocide are scarce. However, river pollution, especially the urban river pollution, is not anything new, but is rapidly increasing. For instance, a recent study highlighted that the

Dhaleshwari River would be under threat due to the relocation of the tannery industries (Farhin 2017:March 29). Another study showed that the Dhaleshwari River is already heavily polluted and vulnerable to further pollution (Ahmed *et al.* 2018:1–4). Reportedly, 1,500 fishermen have been forced to change their livelihood options as Dhaleshwari is no longer sustainable for marine lives. The report also highlighted that the river has become smelly, and its color has changed after few months of the tannery relocation. Let alone for drinking purposes, the water of Dhaleshwari cannot even be utilized for non-drinking household use and agricultural purposes (*The Daily Star* 2019:August 4).

In 2003, a plan had been adopted to shift the tannery industry from Hazaribagh to Savar (Kashem and Islam 2021:July 18), and in 2017, only 35% of the tanneries have been shifted to Savar Tannery Estate (*The Daily Star* 2017:June 11). However, this tannery estate is still incomplete since there is no solid waste management system and central effluent treatment plant (CETP) (Kashem and Islam 2021:July 18). Although this relocation intended to save Dhaka and Buriganga River from excessive pollution, the tannery estate is now threatening the Dhaleshwari River. There are very few past studies that focus on the dynamics of tannery relocation and the pollution of the Dhaleshwari River. Moreover, no study has been recorded that assesses Dhaleshwari River's pollution as ecocide and probes into the political-economic factors. The effect of pollution in the Dhaleshwari River on people's livelihood and the socio-environmental sustainability of the area remained almost absent in previous researches. Furthermore, there are very few past researches in the context of the tannery industry of Bangladesh that focus on climate change as a result of ecocides that the tannery industry causes.

In this chapter, the authors attempt to link two critical aspects together, ecocide and political-economic factors, to assess the alarming issue of urban river pollution from a unique lens. The tannery industry of Bangladesh is significant for the country's economic growth. However, the tannery industry has already destroyed the Buriganga River and now threatening the Dhaleshwari River. Such detrimental practices can easily elevate the impact of climate change. This study also aims to explore the effect of such occurrence from the perspective of livelihood and socio-economic sustainability, highlighting the existing problems in a systematized way.

Research Objective

This research explores the political-economic factors behind the ecocide in the Dhaleshwari River done by the tannery factories in Savar.

Furthermore, this study attempts to probe into the sustainability of the people's livelihood and other socio-economic factors dependent on the ecology of the Dhaleshwari River. This chapter also scrutinizes the impact of ecocide on climate change in the long run in a country like Bangladesh, which is one of the biggest victims. Following are the research objectives that steer the study:

(a) To scrutinize the political and economic forces behind ecocide in Dhaleshwari River due to tannery industry pollution.
(b) To evaluate if ecocide in the Dhaleshwari river is creating an impact on the livelihoods of local people and affecting the socio-environmental sustainability of the people living in the adjacent area.
(c) To review the long-term impact of the river ecocide from the perspective of climate change.

Conceptual Framework

The conceptual framework of this study is inspired by the Elite Theory (Higley 2010:161–176) that explains the power dynamics. According to the theory, a small group of elite people hold most of the power due to their economic supremacy and social capital. In this study, the focus has been set on the political-economic factors controlled mainly by the elite group (such as tannery owners, policymakers, civil society, etc.). The conceptual framework (Figure 1) highlights how the lack of proper socio-political intervention has been causing ecocide in the Dhaleshwari River.

The framework also shows that the current prevalence of ecocide is affecting the livelihood of local people and hampering the socio-environmental sustainability of the area. Resultantly, such occurrence of ecocide might develop a new form of political-economic context. Since the comparatively less powerful people (the marginalized ones) are losing their power, the new form of political-economic circumstance will serve the existing elite group. Assumedly, this sort of shift in the power dynamics will embolden the ecocide that might cause a surge in the effects of climate change in the long run.

Methodology

This qualitative research is planned following the phenomenological research design (Lester 1999:1–4). This research method prioritizes the

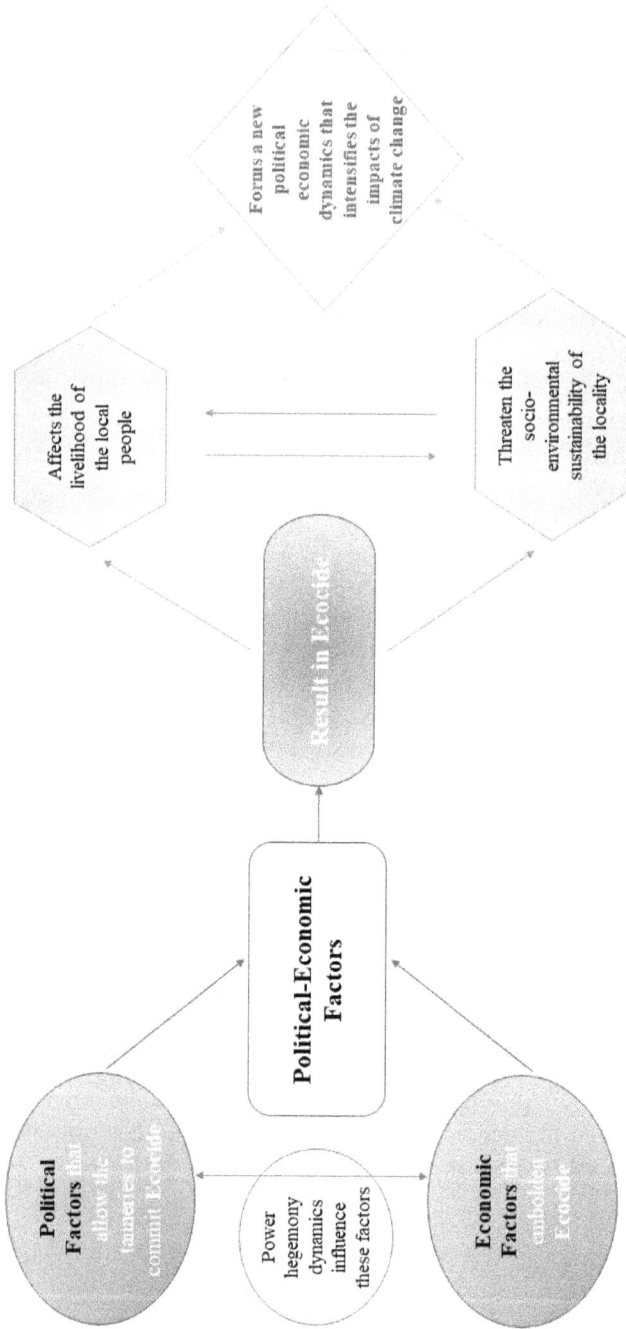

Figure 1: Conceptual framework shows how ecocide has been being influenced by political-economic factors that might augment the effects of climate change in the future.

Table 1: Categorization of interviewed people in the study.

Elites who possess political power	Elites who own economic power	Probable victims of ecocide	Activists/support groups
Bangladesh Tanners Association (BTA)	Tannery owners	Previous fishermen	Bangladesh Environmental Lawyers Association (BELA)
Bangladesh Small and Cottage Industries Corporation (BSCIC)		Residents	Civil society (academics, researchers, etc.)
Local representatives (Union Parishad)		Local youth group	

reality of the stakeholders' experiences. Following the convenience sampling method (Stratton 2021:373–374), 1 Focus Group Discussion (FGD) with the past fishermen who have had to change their profession, 1 FGD with the residents, 1 FGD with the local youth group, 1 Key Informant Interview (KII) with the local representative, 2 KIIs with tannery owners, 1 KII with the representative of Bangladesh Tanners Association (BTA), 1 KII with the Project Director of BSCIC (Bangladesh Small and Cottage Industries Corporation) Tannery Industrial Estate, 1 KII with the focal person of Bangladesh Environmental Lawyers Association (BELA), 1 KII with an Assistant Professor of Institute of Leather Engineering & Technology (ILET), University of Dhaka, and 2 KIIs with researchers who work in the Tannery industry of Bangladesh have been conducted. The qualitative research methodology has been adopted in this study since this tactic involves an interpretive and naturalistic approach (Denzin and Lincoln 2005:22). Table 1 highlights how the data collection procedure attempts to engage different elites from the different strata of the power dynamics.

The interpretive phenomenological analysis (IPA) methodology (Smith and Osborn 2015:41–42) has been followed to analyze the collected data. The study's data collection has been done from August to September 2021, following a hybrid method (both virtual and in-person).

Area of Study

The study has been conducted in Hemayetpur, Savar, Dhaka, where the BSCIC Tannery Industrial Estate is located. The river Dhaleshwari is situated just beside this industrial establishment (Figure 2).

Figure 2:   BSCIC Tannery Industrial Estate and Dhaleshwari River on Google Earth.

Hemayetpur is under the jurisdiction of No. 8 Tetuljhora Union of Savar Upazila. According to the government website,** Tetuljhora Union is 18.05 square kilometers with a total of 20,000 (approximate) voters.

This research has also reviewed journal articles, newspaper articles, books, and other sources focusing on the Dhaleshwari River and the impact of tannery effluent on the river. In terms of assessing the secondary data, the timeframe of the last 5 years (2017–2021) has been considered.

Findings

The political-economic dynamics of ecocide in Dhaleshwari River

From the KIIs, it has been found that some new factories have mush-roomed on the bank of the Dhaleshwari River, centering on the tannery business. Since BSCIC only regulates the registered tannery industries, formal regulation does not necessarily include those rising factories. Such factory owners with comparably influential economic power can manipu-late environmental rights and justice. If they start to exercise their control, it might be challenging to ensure sustainable development there.

Besides, the regulated tannery industries adopt CETP to manage wastewater produced by the industries. Additionally, solid waste is also generated by the factories. Therefore, solid waste management is not envi-ronmentally friendly as they are mostly dumped into the Dhaleshwari River. Besides, the capacity of the CETP is not enough to maintain the waste produced by the factories. The situation worsens, especially during the Eid ul-Adha, when a massive volume of raw material of leather indus-tries (i.e., animal hides) is managed from the sacrificed animals during the festival. Such pollution endangers the Dhaleshwari River and its aquatic organisms.

The tannery industries in the Dhaleshwari River have been allotted 155 plots by BSCIC. Without a doubt, this is a great advantage received by the factory owners. In contrast, the socio-economic condition of the local people has been adversely altered. Several marginal fishermen have lost their livelihoods because the availability of fish is considerably shrinking. The factories' footprint is deteriorating the natural water quality

---

** http://savar.dhaka.gov.bd/.

of the river that is required to sustain the fish. The river water is not usable for other domestic purposes. Local people have raised their voices against the waste dumping factories. However, adequate steps have not been taken by the authorities to resolve this problem. There is a power imbalance between ordinary local people and comparably affluent factory owners. Therefore, this issue augments negative control of natural resources and public property like the Dhaleshwari River. The financial strength makes factory owners influential in decision-making regarding the use of the Dhaleshwari River. Nevertheless, such benefit comes at the cost of the livelihoods of the impoverished fishermen and other local dependents on the river.

The industrial pollution by the tannery industries has indeed led to an adverse situation for the Dhaleshwari River. There are environmental laws (Environmental Pollution Control Ordinance 1977, Bangladesh Environment Conservation Act, 1995, the EIA Guidelines for Industries 2021, etc.) and compliance criteria for waste management for industrial operations in Bangladesh. Despite laws and regulations, tannery industries can continue discharging mismanaged industrial effluents into the water. The economic cost of managing waste is also interlinked with this issue. In the meantime, the local people whose livelihood depends on the river ecosystem are socially and economically exclusive. Displacement of their livelihood increases their vulnerability.

The impact of ecocide in Dhaleshwari River: Affecting the people's livelihood and the socio-environmental sustainability

The ongoing ecocide in the Dhaleshwari River because of industrialization seriously impacts the environment and people's livelihood. The tannery effluent and the untreated solid waste dumped into the river do not only pollute the river but this river ecocide also affects the overall socio-environmental sustainability of the area (i.e., Hemayetpur, Savar). Therefore, the health condition of the local people, social acceptance, and dignity of the area are devalued. The same situation has also prevailed in the former tannery hub where several pollutants such as Chromium, Sulphur, Ammonia, and salt were found. A recent article in a national daily stated that due to the lack of a proper and functioning CETP, 15,000 cubic meters of untreated waste are dumped into the Dhaleshwari River each day (*Dhaka Tribune* 2021:September 11). As a result, the biodiversity and ecosystem of the river and land surrounding the area are being hampered.

The industrial transition near the river Dhaleshwari has affected the livelihood of the local people reducing the fishing activity drastically. The fishing community has been forced to seek alternative income sources. Although few of the affected local people have been able to find a job in the tannery factories, majority failed due to inadequate skillset and experience. Resultantly, most of them had to resort to alternative income sources such as pulling rickshaws, selling utilities, etc. Findings from the FGD reflected that the local people were hopeful about relocating the tannery industry in Savar. They were of the view that these big companies would establish better schools, playgrounds, and other improved facilities for the locality. In reality, the scenario is entirely different. A representative of BELA has informed that the factories can try to utilize their corporate social responsibility (CSR) fund to mitigate these challenges. However, the factories do not take up any such interventions.

Additionally, a significant portion of the local population had to migrate elsewhere to seek livelihood. However, the impact is not limited to that only. The local dwellers near the tannery industrial zone have been the victim of health hazards. Because of severe odor pollution, a significant number of the local dwellers are suffering from several respiratory issues and skin diseases. Moreover, they have to spend a lot of money on healthcare facilities. The tannery odor also affects the air quality which they breathe. However, during the transition of the tannery industry, they were promised health benefits and other social benefits, which are also non-existent, just like an effective CETP.

Furthermore, the impact of ecocide is beyond affecting the river. One of the previous articles that have been reviewed as a secondary data source shows that the Dissolved Oxygen (DO) content of the Dhaleshwari River is way below the critical level of four milligrams per liter and this is a threat to aquatic organisms (Akter *et al.* 2019:160). The interviewed academician has expressed that they have been teaching sustainable ways of leather production. However, most tannery owners only emphasize monetary benefit, and the tannery workers are not well aware of the negative impacts of polluting the river and nature. The representative of Tetuljhora Union Parishad echoed the concerns of local people. He mentioned that this industry relocation was a matter of hope and joy for them; rather, it has turned into a nightmare. The Union Parishad has minimal power to tackle this issue. The majority of the young people who have participated in the FGD do not want to stay in this place if they are given an option. Finally, the research team has found strong evidence that the ecocide has been affecting the socio-environmental sustainability of the area. Since there is no mitigation measure in place, the ecocide is also influencing climate change which places environmental sustainability of Savar under threat. The social sustainability of the place is highly affected since the ecocide obstructs livelihood, sound growth, and dignity of a significant number of people.

Ecocide in Dhaleshwari River: Exacerbating the impacts of climate change

Dhaleshwari is not the first river that has been affected by ecocide. Buriganga was the previous victim of the tannery industry. The river eco-system of the country is highly significant to tackle the issue of climate change. FGD with the local people has reflected that the agriculture pattern has already been changing in Savar. Ecocide in Dhaleshwari is undoubtedly playing one of the most prominent roles here. One of the researchers interviewed has referred to a recent newspaper article that mentioned that The Environment, Forest and Climate Change Ministry has already suggested shutting down the tannery estate operations. He has also informed that the tannery estate does not have the facility to treat solid waste that contains heavy metals. The secondary data reviewed also vouched for this statement. Interestingly, the Department of Environment did not provide clearance to this estate (*The Daily Star* 2021:September 12). However, the BSCIC focal person did not emphasize this issue and underlined that the tannery industry is not harming the environment.

Discussion

Sustainable leather production: Lessons from neighboring countries

According to Gombault and Begeer (2013:57–60), in some leather industries in Brazil, disposal of untreated waste pollutes the river Passo Fundo and causes the death of fish in the river areas. The local small farmers, in general, get little access to the market and sometimes they work for the large slaughterhouse owners at a minimal salary. The report also showed the scenario of China, where in places like Xinji city, local people suffer from the activities of the leather industries that contaminate the drinking water (Gombault and Begeer 2013:49–50).

Similarly, in India, industrial effluent from the tanneries is frequently dumped in rivers. In Southern India, the Palar River is one of the victims in such a case (Gombault and Begeer 2013:43–44). In Vietnam, one of the most pollutant-causing industries is the tanning industry which often aggravates environmental problems by discharging waste without proper treatment (Gombault and Begeer 2013:51–52).

One significant issue in Bangladesh is that the local people and fishermen who belong to the lower-income category have less power to raise their voices than their wealthy counterparts, like tannery owners. Whereas the natural resources are supposed to be distributed equally, the factory owners have a greater chance to alter the natural resources. Besides, the political and decision-making involvement of ordinary people is inferior. Therefore, political and economic factors may be a barrier to ensuring inclusive rights and environmental justice to extract benefits from natural resources. The economically powerful factory owners can advance their operation with existing laws and regulations, while the fishermen have to switch their profession and embrace unwanted socio-economic vulnerability. Consequently, the uneven power distribution and an antagonistic economic motive are crucially responsible for environmental pollution and the violation of environmental justice.

The practice of ecocide around the globe

In April 2010, Polly Higgins, a Scottish lawyer, placed a proposal to the UN Law Commission where she pleaded to include "ecocide" as the fifth crime against peace (Greene 2019:2). Establishing ecocide as a crime is not a new practice across the globe. Countries like Armenia, Belarus, Ecuador, Georgia, Kazakhstan, Kyrgyzstan, Moldova, Russia, Tajikistan, Ukraine, Uzbekistan, and Vietnam have marked ecocide as a crime (García 2020:February 27). The mentioned countries worldwide have labeled ecocide a crime and enacted a law regarding that, and included it in their national constitutions. Russia in Article 358 of the criminal code, Kyrgyzstan in Article 374, Tajikistan in Article 400, Belarus in Article 131, Georgia in Article 409, Vietnam in Article 342, Ukraine in Article 441, Moldova in Article 136, Armenia in Article 394, Kazakhstan in Article 169 of their respective criminal codes have labeled ecocide as a crime (Lamas 2017:16–17). While the global legal-environmental experts are trying to make "ecocide" a crime that is punishable by International Criminal Court (ICC) (Deursen and Eggens 2021:June 1), Bangladesh is also considering labeling ecocide as a crime (*The Financial Express* 2021:June 21).

The initiative to establish ecocide as a crime took a new turn with Ecuador's dedication to an entire chapter in its constitution in 2008 to protect nature (Southwood 2015:6). Bolivia adopted the law of the rights of the mother earth in 2010 (Buxton n.d.), and finally, the United States of America passed an ordinance in 2013 which included rights of nature (Kauffman and Martin 2018:46). Countries such as New Zealand, Spain, and India also took an extensive initiative (Mehta and Merz 2015:6). NGOs in countries such as Belgium, France, and the Philippines took the initiative to look for climate justice through human rights mechanisms (Wijdekop 2016:1). The practice to stop ecocide took a revolutionary turn with the European Citizen Initiative (ECI). The ECI was launched to criminalize ecocide and ensure accountability, prohibiting any form of ecocide in the European territories and calling out for a sustainable economy (European Citizen's Initiative 2012:3).

Industrialization, climate change, and ecocide

A recent study by Wadanambi *et al.* (2020:88) showed that the Industrial Revolution inspired human beings to migrate to urban areas. Resultantly, the urban environment has been negatively impacted. In this context, the concept of ecocide has remained silent and unheard for a long time. Over the world, industrialization did not stay ecofriendly. Industrial waste is one of the significant contributors to climate change (Mgbemene *et al.* 2016:302). The leather industry has been one of the biggest villains since it highly relies on animal farming (Memedovic and Mattila 2008:483). Moreover, excessive use of chemicals is another trait of this industry. Being a 90-billion-dollar industry, the monetary benefit has always been the leather industry (Memedovic and Mattila 2008:485). Sustainable industrialization has always been a concern. Resultantly, this industry has dramatically stirred climate change.

The process and the wastes and effluents from the tanneries are severely toxic and difficult to manage. The leather industry consumes a massive amount of water (and causing river ecocide in Bangladesh).

Moreover, leather has the most significant impact on eutrophication, a severe ecological problem (Somody 2021:July 12). Eutrophication, the overgrowth of plants in the water, is caused by waste that kills other animals by depleting oxygen levels in the water. Estimated, animal agriculture is responsible for more greenhouse gases than all of the world's transportation systems combined (Cameron and Cameron 2017:December 4). A large amount of fossil fuel is required during live-stock production. Hence, the leather industry has an environmental impact (almost three times) than other synthetic alternatives (Somody 2021:July 12). The leather industry does not only harm the animals but also it indirectly (yet immensely) the human being (by attacking the ecosystem). The leather industry is not any by-product of the meat industry, and processing hides into leather is not waste management. Instead, it is an unsustainable but profitable business.

According to a study, leather's carbon footprint ranges between 65 and 150 kilograms of carbon dioxide (CO_2) per m² of leather (Chen *et al.* 2014:1063–1066). The study has also mentioned that each year almost 2 billion square meters of leather is produced. So, the contribution of the leather industry toward climate change cannot be ignored at all. In the developing countries (which are also prone to climate change), environmental impacts of the tannery industries are the outcomes of irresponsible production facilities and illegal waste dumping most of the time. In India, one of the major exporters of leather, 1 m² of hide produces 16,500 liters of wastewater that contains Chromium, Sulfates, and pathogens (Somody 2021:July 12). Such ill practice harms the environment and hinders the health condition of the tannery workers and the local people. In a country like Bangladesh, industrial activities must be brought under proper regulation to tackle climate change effectively. Moreover, if the current trend of the tannery factories is labeled as ecocide and addressed legally, it will provide Bangladesh protection from the adverse impacts of climate change. Sustainable industrialization, responsible consumption, and plant-based leather alternatives can be worthwhile options in this regard.

Recommendations and way forward

The above discussion points to the ramifications of the leather industry on river water. The following recommendations are provided as the way forward.

Responsible production approach and inclusive operational strategy

One of the primary rationales behind shifting the tannery hub from Hazaribagh, Dhaka to the BSCIC Industrial Estate in Savar was that a CETP would control the waste management. In reality, that did not take place. Since the leather industry is one of the biggest contributors to the GDP of Bangladesh, the companies and factories in the estate earn a significant amount of profit each year. Assumedly, they must have a CSR fund. If these factories utilize their CSR funds responsibly to address the damage done by them, then it will significantly reduce the problems in the short run. Utilizing these funds, they can address the issue of livelihood, health, and social issues.

Moreover, tannery estate must have an operational guideline. The operational procedure needs to be updated to ensure proper and effective implementation of the laws. The upgradation must be done collaboratively and inclusively where all the local people (especially, who suffer from the ecocide) will have a voice. Following such accountable procedures will check ecocide in the long run.

Developing a legal framework

There is room for change in terms of sustainability. However, since many commercially significant factories are already conducting their production activities in the BSCIC tannery estate, it might be complicated to replace them. Nevertheless, strict laws should be proposed and implemented for those current factories; particularly, laws concerning "ecocide" can be incorporated. Furthermore, the law enforcement sector should strictly monitor and handle any infringement of the law. Without such a law, accountability will be less likely to be achieved. Similarly, the influence of the elite and their political power will continue to prolong discrimination.

Additionally, the establishment of factories on the bank of any river should be restrained in future. The condition of the rivers is already dangerous. In the meantime, Bangladesh is a potential victim of climate change. Therefore, to fight climate change, in the long run, ecocide law can be proposed to deal with short-term barriers against sustainable development.

Conclusion

Ecocide seems to be comparatively a new concept, especially in the developing countries. However, it might become easier to deal with complicated socio-political and political-economic factors if the concept of ecocide can be established within the legal framework of a country. Generally, people with power can influence the decision-making process regarding industrialization and its impact on the environment, which ultimately results in unfavorable footprints on natural resources. Eventually, such discrimination and environmental injustice threatens the likelihood of exacerbating climate change in the long run. If not halted, ecocide will keep questioning the approach and sustainability of the development process in Bangladesh. Being exceptionally vulnerable to climate change, Bangladesh will experience an even accelerated and intensified climate disaster if such ecocide occurs. To resolve such threats against sustainability, inclusive "ecocide" policy can be undertaken which will safeguard the environment and natural resources and help mitigate the effects of climate change in the long run.

References

Agrawala, S., Ota, T., Ahmed, A. U., Smith, J. and van Aalst, M. (2003). *Development and Climate Change in Bangladesh: Focus on Coastal Flooding and the Sundarbans*. Paris: Organisation for Economic Co-operation and Development.

Ahmed, K. S., Sarkar, A. M., Hossain, H. and Saha, B. (2018). Pollution of Dhaleshwari River due to newly shifted tannery industry in Savar, Dhaka, Bangladesh. *Conference of Young Scientist on the Topic "Workshop on Climate Change"*. Devecha Centre for Climate Change, Indian Institute of Science (IISc).

Akash, M. (2021, July 7). Safari park in Lathitila forest: Are we committing ecocide? *The Business Standard*. https://www.tbsnews.net/environment/nature/safari-park-lathitila-forest-are-we-committing-ecocide-271360.

Akter, S., Kamrujjaman, I. and Saha, B. (2019). An investigation into chemical parameters of water of Dhaleswari: A river alongside tannery village of Bangladesh. *International Journal of Science*, 8, 159–164.

Buxton, N. (n.d.). The law of mother earth: Behind Bolivia's historic bill. *Global Alliance for the Rights of Nature*. https://www.therightsofnature.org/bolivia-law-of-mother-earth/.

Cameron, J. and Cameron, S. A. (2017, December 4). Animal agriculture is choking the Earth and making us sick: We must act now. *The Guardian*.

https://www.theguardian.com/commentisfree/2017/dec/04/animal-agriculture-choking-earth-making-sick-climate-food-environmental-impact-james-cameron-suzy-amis-cameron.

Chang, C., Benson, M. and Fam, M. M. (2017). A review of Agent Orange and its associated oncologic risk of genitourinary cancers. *Urologic Oncology: Seminars and Original Investigations, 35*(11), 633–639.

Chen, K., Lin, L. and Lee, W. (2014). Analysing the carbon footprint of the finished bovine leather: A case study of aniline leather. *Energy Procedia, 61,* 1063–1066. https://doi.org/10.1016/j.egypro.2014.11.1023.

David, W. (2020). *Ecocide: Kill the Corporation before It Kills Us.* Manchester: Manchester University Press.

Denzin, N. K. and Lincoln, Y. S. (2005). *The SAGE Handbook of Qualitative Research* (3rd ed.). New York: Sage Publications, Inc.

Deursen, J. V. and Eggens, N. (2021, June 1). Environmental crimes in international law: An exploration of the possibilities for criminalisation of ecocide. *Deviance Incubator.* https://devianceincubator.wordpress.com/2021/06/01/environmental-crimes-in-international-law/.

Dhaka Tribune (2021, September 11). DoE to BSCIC: Why should Savar tannery estate not be shut down? *Dhaka Tribune.* https://www.dhakatribune.com/bangladesh/dhaka/2021/09/11/doe-to-bscic-why-should-savar-tannery-estate-not-be-shut-down.

European Citizens' Initiative (2012). *End Ecocide in Europe: A Citizen's Initiative to Give the Earth Rights.* http://www.endecocide.org/wp-content/uploads/2013/01/documents/End-Ecocide-in-Europe-A-Citizens-Initiative-to-give-the-Earth-Rights.pdf.

Farhin, N. (2017, March 29). Dhaleshwari River under threat of pollution. *Dhaka Tribune.* https://www.dhakatribune.com/bangladesh/2017/03/29/dhaleshwari-river-threat-pollution.

García, B. (2020, February 27). The history of ecocide, a new crime against humanity. *OpenMind.* https://www.bbvaopenmind.com/en/science/environment/the-history-of-ecocide-a-new-crime-against-humanity/.

Gombault, M. and Begeer, A. (2013). *Sustainability in the Leather Supply Chain.* Utrecht: MVO Netherlands.

Greene, A. (2019). The campaign to make ecocide an international crime: Quixotic quest or moral imperative? *Fordham Environmental Law Review, 30*(3), 1–49.

Higley, J. (2010). Elite theory and elites. In K. Leicht and J. Jenkins, *Handbook of Politics: Handbooks of Sociology and Social Research,* pp. 161–176. Berlin: Springer.

Kashem, A. and Islam, R. (2021, July 18). Savar Tannery Estate: Complete yet incomplete after 19 years. *The Business Standard.* https://www.tbsnews.net/economy/savar-tannery-done-much-be-done-after-19-years-276577.

Kauffman, C. M. and Martin, P. L. (2018). Constructing rights of nature norms in the US, Ecuador, and New Zealand. *Global Environmental Politics, 18*(4), 43–62. https://doi.org/10.1162/glep_a_00481.

Lamas, C. A. (2017). Ecocide addressing the large-scale impairment of the environment and human rights (Unpublished master's thesis). Global Campus Open Knowledge Repository.

Lester, S. (1999). *An Introduction to Phenomenological Research.* Taunton: Stan Lester Developments.

Loodin, N. (2020). Aral Sea: An environmental disaster in twentieth century in Central Asia. *Modeling Earth Systems and Environment, 6*(4), 2495–2503. https://doi.org/10.1007/s40808-020-00837-3.

Luetz, J. (2017). Climate change and migration in Bangladesh: Empirically derived lessons and opportunities for policymakers and practitioners. *Climate Change Management,* 59–105. https://doi.org/10.1007/978-3-319-64599-5_5.

Maskun, Achmad, Naswar, Assidiq, H. and Bachril, S. (2021). Palm oil cultivation on peatlands and its impact on increasing Indonesia's greenhouse gas emissions. *IOP Conference Series: Earth and Environmental Science, 724*(1), 012092. https://doi.org/10.1088/1755-1315/724/1/012092.

Mehta, S. and Merz, P. (2015). Ecocide — A new crime against peace? *Environmental Law Review, 17*(1), 3–7.

Memedovic, O. and Mattila, H. (2008). The global leather value chain: The industries, the main actors and prospects for upgrading in LDCs. *International Journal of Technological Learning, Innovation and Development, 1*(4), 482–519.

Mgbemene, C. A., Nnaji C. C. and Nwozor, C. (2016). Industrialisation and its backlash: Focus on climate change and its consequences. *Journal of Environmental Science and Technology, 9,* 301–316. https://doi.org/10.3923/jest.2016.301.316.

Ministry of Foreign Affairs of the Netherlands (2018, April). Climate change profile Bangladesh. www.government.nl/foreign-policy-evaluations.

Naser, M., Swapan, M., Ahsan, R., Afroz, T. and Ahmed, S. (2019). Climate change, migration and human rights in Bangladesh: Perspectives on governance. *Asia Pacific Viewpoint, 60*(2), 175–190. https://doi.org/10.1111/apv.12236.

Onyena, A. P. and Sam, K. (2020). A review of the threat of oil exploitation to mangrove ecosystem: Insights from Niger Delta, Nigeria. *Global Ecology and Conservation, 22,* e00961. https://doi.org/10.1016/j.gecco.2020.e00961.

Raftopoulos, M. and Morley, J. (2020). Ecocide in the Amazon: The contested politics of environmental rights in Brazil. *The International Journal of Human Rights, 24*(10), 1616–1641. https://doi.org/10.1080/13642987.2020.1746648.

Rahman, M. (2018). Governance matters: Climate change, corruption, and livelihoods in Bangladesh. *Climatic Change, 147*, 313–326. https://doi.org/10.1007/s10584-018-2139-9.

Sachs, J., Schmidt-Traub, G., Kroll, C., Lafortune, G., Fuller, G. and Woelm, F. (2020). *The Sustainable Development Goals and COVID-19: Sustainable Development Report 2020.* Cambridge: Cambridge University Press.

Sayeda, U. (2021, April 17). Ecocidal displacements in Bangladesh: A climate-affected hotline zone. *Act for Displaced.* https://actfordisplaced.org/2021/04/17/ecocidal-displacements-in-bangladesh-a-climate-affected-hotline-zone/.

Schwegler, V. (2017). The disposable nature: The case of ecocide and corporate accountability. *Amsterdam Law Forum, 9*(3), 71–99. https://doi.org/10.37974/ALF.307.

Smith, J. A. and Osborn, M. (2015). Interpretative phenomenological analysis as a useful methodology for research on the lived experience of pain. *British Journal of Pain, 9*(1), 41–42.

Somody, F. (2021, July 12). Why does the leather industry need to change? *Leap.* https://www.explore-leap.com/post/why-the-leather-industry-needs-to-change.

Southwood, J. (2015, April 20). So what? A critical analysis of the 'greening' of Ecuador's constitution (Unpublished PIED dissertation). Leeds: The University of Leeds.

Stratton, S. J. (2021). Population research: Convenience sampling strategies. *Prehospital and Disaster Medicine, 36*(4), 373–374.

The Daily Star (2017, June 11). 35% tanneries shifted to Savar. *The Daily Star.* https://www.thedailystar.net/city/35-tanneries-shifted-savar-1418707.

The Daily Star (2019, August 4). Tanneries killing the Dhaleshwari River. *The Daily Star.* https://www.thedailystar.net/editorial/news/tanneries-killing-the-dhaleshwari-river-1781302.

The Daily Star (2021, September 12). Savar tannery complex: Environment ministry taking steps as per parliamentary committee's recommendation. *The Daily Star.* https://www.thedailystar.net/environment/news/savar-tannery-complex-environment-ministry-taking-steps-parliamentary-committees-recommendation-2174236.

The Financial Express (2021, June 21). Parliamentary body suggests codifying ecocide as crime like genocide. *The Financial Express.* https://thefinancialexpress.com.bd/national/parliamentary-body-suggests-codifying-ecocide-as-crime-like-genocide-1624276268.

Wadanambi, R. T., Wandana, L. S., Chathumini, K. K. G. L., Dassanayake, N. P., Preethika, D. D. P. and Arachchige, U. S. P. R. (2020). The effects of industrialisation on climate change. *Journal of Research Technology and Engineering, 1*(4), 86–94.

Wijdekop, F. (2016). Against ecocide: Legal protection for earth. *Great Transition Initiative*. https://www.greattransition.org/publication/against-ecocide.

Zepeda, R. (2017). *Human Rights and Environmental Impacts of Palm Oil in Sayaxche, Guatemala.* London: Oxfam.

Zierler, D. (2011). *The Invention of Ecocide: Agent Orange, Vietnam, and the Scientists Who Changed the Way We Think About the Environment.* Georgia: University of Georgia Press.

Chapter 6

Challenges of Space Debris and Space Drag: Building an International Climate Change Regime

Swasti Rao* and Kunwar Alkendra Pratap Singh*

Abstract

Climate change poses one of the most tangible security challenges of our times. The effort to securitize the issue has sped up since the last decade owing to its increasing existential threats. In spite of this growing realization and development of international regime complex for climate change (Jervis 2020:21), the issue still suffers from what is called a "Climate Change Securitisation Paradox" (Lucke *et al.* 2020). On the one hand are several securitization efforts by activists, government, and trans-governmental actors revving up a national security debate around it, but on the other hand there remains lack of consensus to undertake much needed extraordinary measures to tackle this inordinate threat. This disjunct between narratives and praxis is even more glaring in outer space where major players have been expanding their futuristic power

*Swasti Rao is Associate Fellow at Manohar Parrikar Institute for Defence Studies and Analyses (New Delhi) and formerly Assistant Professor at Department of Strategic and Security Studies, Aligarh Muslim University (Aligarh) and Dr Kunwar Alkendra Pratap Singh is Assistant Professor at Department of Physics at Banaras Hindu University, Varanasi.

maximization capabilities. What makes research on climate governance in space daunting and yet indispensable is the inevitability of the space age in human life. It is indeed no exaggeration to state that whoever controls space, will control the future. However, we must manage the spillovers from the space race, most notably in how they could potentially alter climate considerations that may threaten human life back on Earth. This chapter examines two most prominent factors where competitive spacefaring behavior has complicated the already complex and highly debated issue of climate change. First is the problem of space debris polluting and cluttering the Lower Earth Orbit (LEO), and second is the phenomenon of space drag which is directly related with the greenhouse emission produced on Earth which also adds to space cluttering. The chapter analyzes key events and patterns that have led to the increase in space debris and space drag. It highlights the lack of a strong regime complex to control the repercussions of the irresponsible spacefaring behavior of countries like China. It concludes by proposing ways to strengthen the Regime Complex for Climate Change by delving in the concept of Space 2.0 that emphasizes on techniques of augmentation and reconstitution to make spacefaring more cost effective, reusable, and transparent and thereby lessening the problem of space cluttering.

Keywords: Climate Change, Space, Debris, Space-Drag, Greenhouse Emissions, Spacefaring

Introduction

The planetary health debates contend that the health of the planet is a prerequisite to sustaining human civilization. This is the larger context in which this chapter examines the state of increasing clutter in the Lower Earth Orbit (LEO) and its connections with climate change which has serious implications that have not been sufficiently examined. The last decade has of course seen Anthropocene approach becoming mainstream and for the first time the Paris Agreement has created a legally binding international treaty that brings hope as all the major players agree upon their Nationally Determined Contributions (NDCs). This has clearly strengthened the case for building a comprehensive climate change governance regime and recent years have witnessed far more pragmatic initiatives by major powers in evolving such an overarching paradigm. The next path-breaking event by the 26th COP at Glasgow in November

2021 is that arch rivals U.S. and China affirmed their commitments in meeting their Paris goals (US Department of State 2021). This makes climate change the key area of convergence in a world that is otherwise marred by power rivalries.

Global experience of coronavirus pandemic has only reinforced shared sense of climate challenges as also reinforced focus on space and its linkages to climate change trajectories. The relentless maximization of space capabilities among spacefaring nations, burgeoning space technologies, and applications aimed for the use of outer space, have become an increasingly important and even indispensable factor to reckon to. The overarching utility of outer space has strengthened its federating force compelling countries to coordinate and cooperate with each other for various functional necessities. Space technologies and applications have become vital for a wide range of governing issue areas, therefore space governance as such is no longer limited to regulating the use of outer space, but has implications for the governance systems of other global issues for guaranteeing human safety, for enhancing socio-economic development, and for safeguarding environmental sustainability. These separate global governance systems are becoming closely connected and interconnected with the space regime complex as such because of the broadened utilities of space technologies and applications (Liao 2016).

International Climate Change Regime in Space

Conceptualizing the creation of an international climate change regime in space must begin with focus on two of its salient features that must be understood to know how, to begin with, a space regime is likely to emerge (Liao 2016). In brief these are *dispersion of interests* and *issue-centric attributes*.

The distribution or dispersion of interests refers to how states, by deploying their strategic preferences and influential capacities relevant to the issue area, can lead to the creation of a regime, or change an extant regime to fulfill their self-interest. According to Stephen D. Krasner, national power (interest) is the most ostensible driving force that maintains the power balance in global politics and also the equilibrium of the entire global governance architecture (Krasner 1991:336–366). Likewise in the construction of the space governance architecture too, maximizing national interest emerges as the main driver.

The pertinent question here is what can make this distribution of interest overlap enough to create a regime complex for space? This is better understood by focusing on the functional part of spacefaring issues which affect all parties almost equally back on Earth's surface. The functional necessity equally motivates states to constantly create new regimes, or adjust the functional missions and organizational rules to satisfy their needs and ensure their existence. This brings us to the second characteristic mentioned above — *issue-centric attributes*, which refer to specific issues *per se* that become central to stakeholders at a given time. The pressing concern rising from these issues is what compels to push the stakeholders into rising above competition and align efforts to create sustaining structures that can tackle the issues with efficacy.

(i) *Space Debris*: Logically then, just like politics among nations on Earth, the issues facing the regime complex for space governance has to be multidimensional with elements of cooperation, competition, and conflict. It is in this backdrop that this chapter examines the key areas of climate change and its linkages with issues of space governance. It specifically analyzes two most prominent factors where spacefaring behavior compounds the already complex and highly debated issue of climate change. The foremost among these factors is the issue of Space Debris. This first came into prominence when the first satellite Sputnik was launched in 1957 and debris have been increasing since as more and more nations get into the space race for both military and commercial purposes. Apart from the new entrants, the established ones continue to race against one another to launch satellites and anti-satellites weapons in a bid to outdo each other's posturing especially in the LEO. The problem of space debris polluting the LEO, if uncontrolled, can lead to an existential threat for human life back on Earth as one of the 23,000 cataloged particles (and the billion uncatalogued ones) collision with an existing satellite could lead to uncontrollable effects back on Earth.

(ii) *Space Drag*: The second factor analyzed in this chapter is the phenomenon called the space drag; something which is directly related with the greenhouse emissions produced on Earth. The interesting bit about the greenhouse emissions is that while they increase the temperature on Earth, in outer space they end up doing the exact opposite. Hence, because of a drop in temperature in outer space, they result in a decrease in the density of thousands of debris/junk particles whizzing around the Earth. And this decrease in density makes the objects orbit in space far longer than they would have had there been no space drag. This scenario further clutters the LEO and threatens the risk of collision among objects in the LEO thereby thwarting life on Earth. Together, both these factors, space debris and space drag, have seriously complicated the issue of cluttering the LEO that, if went uncontrolled, threatens to become the formidable existential threat on Earth.

Pandemic-driven Paradigm Shift

In the 21st century, almost all of the world's phenomena occur in open systems, which means that they are created by a variety of causal structures, mechanisms, processes, and fields (Roy 2010). In that sense, the year 2020 has brought a paradigm shift in the ways the world operated because of the pandemic and urged the whole world to take the issue of transboundary challenges in an interconnected world with more gravity and seriousness than ever. It exposed the weaknesses of the global institutions and governance structures to tackle the complex and imminent threat of climate change. This is where the systematic analysis and review of an international regime on climate change comes in. Having said that, the speculation on outer space is still far from satisfactory despite the imminent threat they have for human survival back on Earth. Without understanding this, it is impossible to resolve the paradox mentioned above. The securitization of climate change and strengthening an international regime to tackle its challenges has to be done in both theory and practice.

(i) *Low Earth Orbit*: LEO is a key area which extends to about 2,000 kilometers at 250 miles where the International Space System is orbiting and where mustering of space industries has led to a vast amount of revenue generation to the tune of over $300 billion. After the U.S. Global Positioning System (GPS) and the Russian GLONASS system, the Chinese Beidou Satellite Navigation Systems has placed several satellites in LEO since 2019. As spacefaring nations compete to enhance their power capabilities in LEO and beyond, the dynamics have resulted in increasing space junk (debris to the tune of 100,000 tons) with potential hazards for human survival. Figure 1 shows the

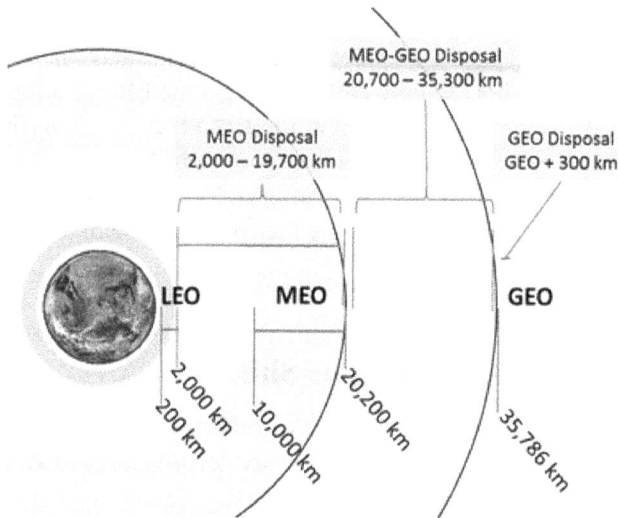

Figure 1: Low Earth Orbit (LEO), Medium Earth Orbit (MEO), and Geosynchronous Orbit (GEO) altitudes (Gheorghe and Yuchnovicz 2015).

expanse of how the Lower, Medium, and Geo-Stationary Earth orbits are divided.

Cluttering of the LEO with space debris/junk has already made it an orbital space junk yard (Orbital Debris Program Office, NASA). Space debris refers to the millions of pieces of space junk flying in LEO most of which comprises human-generated objects, such as pieces of space craft, tiny flecks of paint from a spacecraft, parts of rockets, satellites that are no longer working, or explosions of objects in orbit flying around in space at high speeds (NASA Headquarters Library).

(ii) *Inter-Agency Space Debris Coordination Committee*: The Inter-Agency Space Debris Coordination Committee (IADC) is one of the best known platforms to review the advancement in an international regime complex on climate change on the issue of space debris (IADC 2019). It is an international governmental forum that was founded in 1993 for the worldwide coordination of activities related to the issues of man-made and natural debris in space. The primary purposes of the IADC are to exchange information on space debris research activities between member space agencies, to facilitate opportunities for cooperation in space debris research, to review the

progress of ongoing cooperative activities, and to identify debris mitigation options. In addition to reviewing all ongoing cooperative space
debris research activities between its member organizations, the
IADC recommends new opportunities for cooperation, serves as the
primary means for exchanging information and plans concerning
orbital debris research activities, and identifies and evaluates options
for debris mitigation.

As appropriate, the IADC is expected to communicate the findings
of its work to the wider space community such as the IADC Space
Debris Mitigation Guidelines first published in 2002, and subsequently updated in 2007, 2020, and 2021 (IADC 2021). These IADC
Guidelines are informed, and provide the basis for, the development
of the Space Debris Mitigation Guidelines of the United Nations (UN)
Committee on the Peaceful Uses of Outer Space, which were endorsed
by the UN General Assembly in its resolution 62/217, dated 22
December 2007 (IADC 2017). A Steering Group plus four specified
Working Groups covering measurements (WG1), environment and
database (WG2), protection (WG3) and mitigation (WG4) make up
the IADC (IADC 2019) which has also been active.

(iii) *The China factor*: The IADC ADC, however, has been rendered in
effective by China which has emerged as the new most influential
spacefaring country. In spite of China having become a signatory to
the IADC way back in 1993, and despite the formation of Chinese
National Expert Committee of Space Debris Research in 1999 (Gong
2016), in January 2017, China had gone ahead and performed the
most irresponsible ASAT test ever done in the history of human space
engagement. Expectedly, the Chinese experts, while mentioning the
positive efforts PRC has put into the research on space debris, never
refer to their 2007 debacle (Day 2013). Indeed, a quick glance at most
irrational Chinese military aggression along the LAC with India and
in the Taiwan Straits as also in Taiwanese ADIZ as well as the South
China Sea is also reflected in China's space behavior. The question
then is how to increase the costs for China (or any country that
behaves in a roguish way); or how else to deter such actors from
deflecting from established and agreed norms in space exploration.
The answer lies in creating a strong international regime on space
with a particular emphasis on space debris management. However, the
reliability of that regime is indeed a work in progress that must strive
at improving its vulnerability at the hands of states like China.

The History and Politics of Space Debris

A quantitative overview of types of space debris can be obtained from the General Catalog of Artificial Space Objects (GCAT) (McDowell 2022). Even a very small particle of debris whizzing around the Earth at 18,000 miles per hour can be catastrophic if it hits a space object. Space debris has been around since the very first satellite launch with Sputnik in 1957. The first two pieces of orbital debris were empty R-7 core stage, and nose cone. What is noteworthy is that the oldest piece of debris from those times (NASA 2015) is still in orbit (Gorman 2022). This fact underscores the complexity in the very problem of space debris.

(i) *Mapping cataloged and uncataloged objects*: Total cataloged objects in the LEO are 23,000 but this number comes with a catch. While the cataloge is fairly complete for objects bigger than 10 cm in the LEO, the problem is that it is far from complete at higher altitudes (e.g., in the geostationary orbit (GEO)). The number there could be a million or by some approximations even a billion objects. Out of the 23,000 total cataloged objects, there are 4500 (+−100) Live satellites and everything else is junk. This compels us to look more carefully at the types of debris currently orbiting in the LEO. The major types of Orbital Debris have been classified into (A) Cataloged and (B) Uncataloged objects (Bruno and Perez 2017). Further, the 23,000 cataloged objects in the LEO can be classified as follows:

- Dead satellites — 3,000 cataloged objects.
- Old Rocket Stages — 1,900.
- Littering — 1,700 (Interstage adapters, optics covers, Fairings).
- Disintegration debris that form the largest category with objects running to the tune of 11,800 (comes from exploding rockets, anti-satellite tests, accidental collision, battery explosions, etc.).

Not only from the cataloged but the existential threat posed by the millions of uncatalogued objects in the LEO is a growing concern as well. Uncatalogued debris objects in the LEO are numerically represented as: 0.5 million >1 cm, 0.1–1 billion >1 mm (Bruno and Dolado 2017). In the view of the above, the next ponderable is those specific events that have made the space junk such a menace. A quantitative overview of the major debris events in the history of human intervention in space is discussed next (McDowell 2022). First major debris event was the Explosion of Ablestar 008 Rocket stage that exploded into 300+ Pieces on June 29, 1961. Out of these 300+, 187 objects are still in orbit even 60 years later. Once again that highlights the crux of the problem. The debris objects **REMAIN** in orbits and the number keeps increasing.

(ii) *Milestone events in orbital debris history*: After the Ablestar explosion, events such as the First Upper Stage Disintegration in June 1961, Project West Ford in May 1963, First Anti-satellite test (Kosmos 248/249/252); first military weapon test by USSR in November 1968, First Delta Stage breakup which led to new designs for controlled propellant depletion in December 1973, Kosmos — 954 (US-A-325) nuclear reactor re-entry over Canada in January 1978, Skylab re-entry over Australia in July 1979 — are milestone events that led to design for controlled de-orbiting of large payloads to come back down under control instead of letting them reenter randomly. Following that, a series of events in the 1970s lead the experts to design the rockets differently to allow the engines to restart so that you could use the left over fuel instead of leaving these ticking time bombs. The decade of 1980s saw a new stage of debris creation by battery explosions.

First major battery explosion (Kosmos-1275 ParusNavsat) happened in July 1981 (Bhramand 2009). First major SOZ (Sistema Obespecheniya Zapuska) breakup in high orbit which was a USSR venture in February 1991 that also led to battery explosions (Anz-Meador 2018). However, the worst debris event in the history of human intervention in space came in January 2007. This was the Chinese anti-satellite test which continues to be the largest and long lived deliberate debris event till date (NASA 2021). This was followed by the Iridium 33/Kosmos-2251 collision in February 2009. There are several points to be understood when it comes to discerning patterns in the mapping of problem of space debris/junk. We will take a look at them one by one.

The next section examines two important patterns. The first will analyze the pattern in the growth of space junk that will focus on the parameters of important events that have led to this scenario, the role of China's ASAT, and the difference in cluttering in the Upper and Lower LEO. And the second pattern will analyze the uncontrolled re-entry problem especially in the case of China that has compounded the issue of space junk.

(iii) *Pattern in the growth of space debris*: It is interesting to make a comparison to appreciate what impact any states rogue behavior can make to growth of space debris. Figure 2 gives one most apt

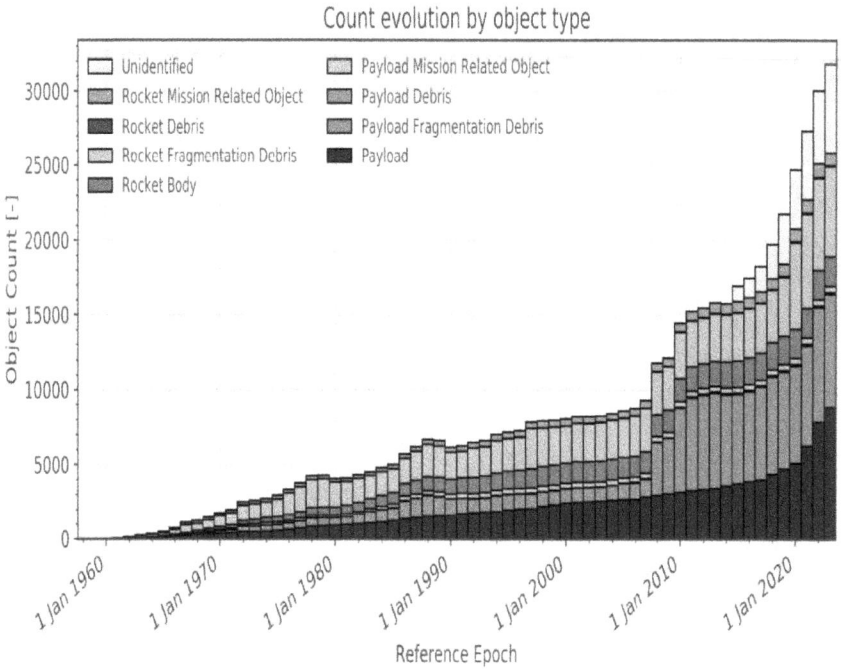

Figure 2:   The growth of space junk.

Source: https://sdup.esoc.esa.int/discosweb/statistics/.

illustration of the growth of Space junk right form the 1960s. The sheer size of different types of junk put together has risen alarmingly.

Compare that to the image given in Figure 3 that shows how China's 2007 Anti-Satellite test was to become a real spoiler in compounding debris. Looking carefully at the portion hogged by the Chinese ASAT's contribution to the growth of space junk by all other spacefaring nations put together definitely raises an alarm. And this alarm becomes a red flag when it is further noted that China exhibited such behavior despite being a signatory to the IADC. This shows how one event by China resulting so significantly in overall increase of space debris also gets compounded by the fact that the overall annual orbital rate for China has gone up alarmingly.

At this stage, it will be useful to throw some light on India's example which has been a responsible spacefaring state. Responsible spacefaring shown by India presents a case study into ways to manage the issue. It is noteworthy that contrary to China's 2007 ASAT debris shock, India carried

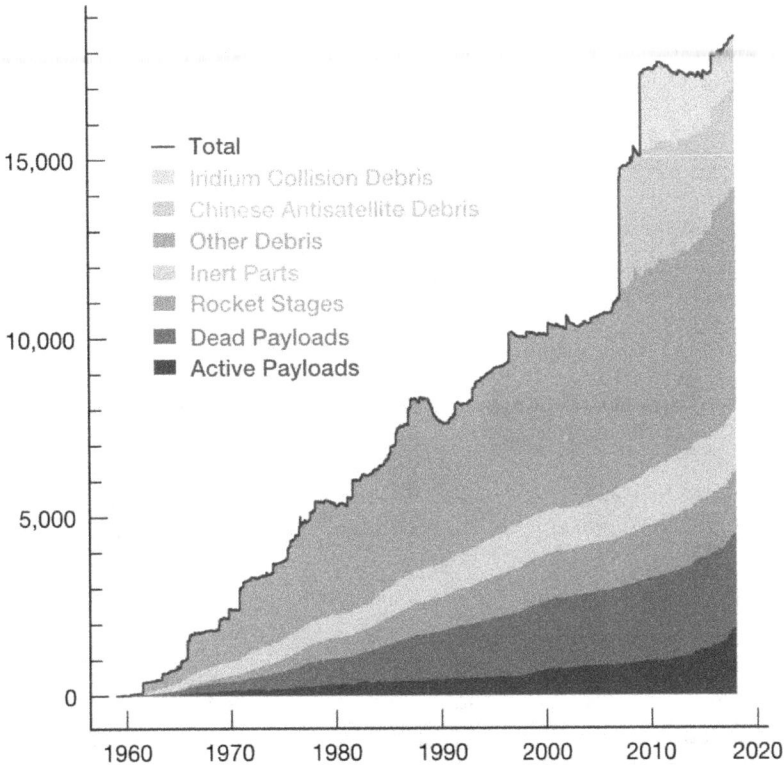

Figure 3: China's ASAT test driven Space Junk.

Source: https://www.planet4589.org/jcm/pubs/space/articles/2018/STEL.debris.pdf.

out a relatively responsible way of testing its ASAT in 2019. The test was done in the lower atmosphere to ensure that there is no space debris. Whatever debris that was generated decayed and fell back onto Earth within weeks (Tellis 2019). Herein emerges the key to understanding a crucial aspect of responsible spacefaring. That space debris will be generated is an unavoidable fact, but if it is generated in a *particular* way, it reduces the chances for that debris to "stay" back in the space orbit to almost negligible proportions. With this, the chance of space accidents is also reduced significantly.

Difference in Lower and Upper in LEO

Rising space junk clutters the LEO in different levels. The LEO (stretches to almost 2,000 km from the surface of the Earth as shown in Figure 1) is

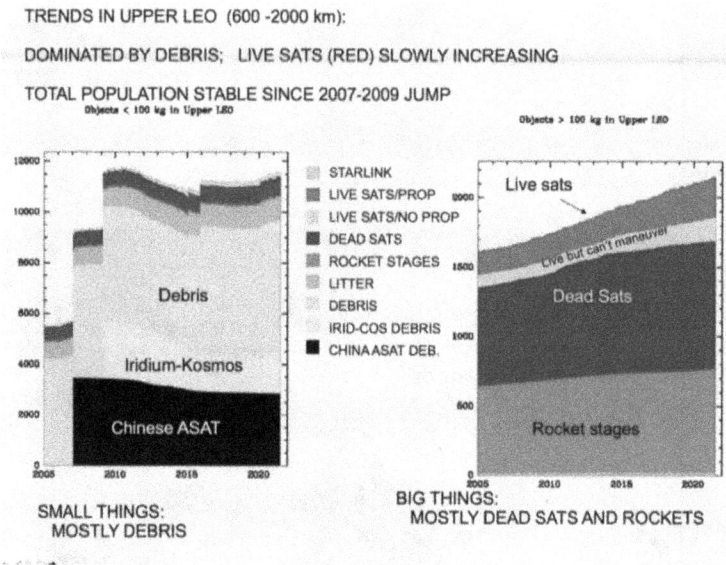

Figure 4:   Trends in Upper LEO from Chinese ASAT.
Source: https://planet4589.org/talks/global/global16.pdf.

not a monolith and can be divided into Lower LEO and Upper LEO with distinct debris trends. Six hundred kilometers is usually assigned as the line that divides lower LEO from upper LEO. First, we will analyze the debris trends in Upper LEO. In last 15 years the small debris has seen a jump with Chinese ASAT in Upper LEO which is illustrated in Figure 4.

However, there is another problem with the debris trend in Lower LEO where a transformative change is being seen in last 5 years. The debris in this region is completely dominated by CubeSats. CubeSats are miniature satellites that have been used exclusively in the lower part of LEO (although now also being used beyond that). Cubesat cluttering is resulting from the deployment of thousands of satellites in the Starlink Internet constellation that Elon Musk's Space X is building. Hence, Lower LEO is dominated not only by debris but also live satellites (see Figure 5). Added cluttering (CubeSats +debris) increases the probability of catastrophic collisions which will be detrimental to space environment. Research shows that over the next 50 years, the trend of deploying CubeSats might result in a formidable increase in the number of collisions (Harris 2014). While mega constellations like Starlinks have increased the number of satellites revolving in Lower LEO, they can still

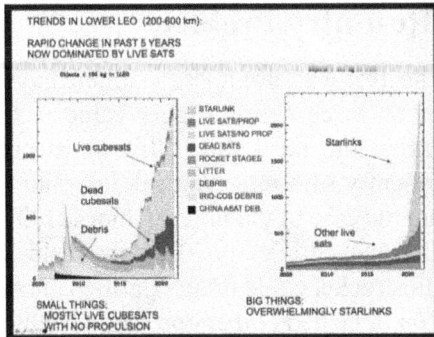

Figure 5: Trends in Lower LEO compounded by CubeSats.
Source: https://planet4589.org/talks/global/global16.pdf.

be carefully managed. It is in the smaller particles that the collision dread resides.

That is not all. The ever increasing problem of space debris has yet another worrying aspect stemming from a phenomenon called the *Van Allen Belts*. These are radiation belts around the Earth and may be understood as rings of charged particle strapped in the Earth's magnetic field. Research has shown how particle radiation environment of these belts works around the Earth. A NASA study showed just how quickly the radiation swings from tepid to extreme (McDowell 2016). This near Earth particle radiation, extremely volatile and random by nature, could cause havoc with the ever-increasing cluttering of LEO. The increasing number of junk and dead Sats in the LEO has already been touching the *Van Allen Belts* (*Ibid.*). The orbital debris is an endemic issue which has been growing since the days of Sputnik in 1957 and growing really fast now. China's ASAT test has been a major spoiler. While the mass of space debris is dominated by dead Sats and dead rockets but the number is dominated by disintegration debris especially in the lower LEO where the change in last five years has been fundamental and alarming.

The problem doesn't end here. We do not even know the situation in higher orbits like the GEO. There is a worrying lack of data there and while Russia has some data on the same, it is classified information so far. The rest of the spacefaring clout headed by NASA is trying hard to rectify this issue. We can only hope that the picture that emerges is not so grim for space environment. Discussed next is the second pattern that increases space junk due to uncontrolled reentries.

Uncontrolled Re-Entry Problem

China's Anti-Satellite Test was also a lesson in one more dimension. Here the culprit for another space debris trigger came in the form of Large Uncontrolled re-entries into the Earth. While most other countries perform a controlled re-entry of a returning satellite into Earth, China does not. Upon examining the case of the famed Chinese LONG march, technically named CZ-5B launched in early 2021, one realizes that the Chinese rocket debris from this rocket could hit the Earth in uncontrolled re-entries ever since it has been launched (Liberatore and Chadwick 2021). And as again, China does not seem to be concerned with this huge problem for space environment. The point to ponder is that even the uncontrolled re-entries, pretty much like their 2007 ASAT could be avoided.

However, the aforementioned Chinese behavior is not isolated. The recent history of how little concern China showed for uncontrolled re-entries will highlight how weak the regime on space has proved so far in dealing with a repeat offender (McDowell 2021). In 2017, China's Taingong space station made an uncontrolled re-entry, ultimately landing in the middle of the Pacific Ocean (Pappas 2018). Next, in May 2020 another Long March 5B rocket made a chaotic re-entry into Earth's atmosphere with some of it landing near the west coast of Africa and pieces of debris fell into two villages in Cote d'Ivoire causing damage to people's business and homes (McDowell 2021).The question is why doesn't China feel any responsibility to build the rocket stage with a booster to steer it into a safe landing point in water only? China's Long March 5B was *not* built with either of those options.

Part of the uncontrolled re-entry problem stems from the fact that unlike the IADC devoted to managing space debris, there are no international regulations that stipulate rockets be built with uncontrolled re-entries mechanisms. It is only a norm followed by space agencies around the world (McDowell 2021). Without necessary and potent international regulations, the fact that China did not consider this in their rocket design reflects badly on the ethics of the country's space program. China's space program is obsessed with stellar performances in landings and explorations; after landing a rover on the Moon in November 2020 and landed their first rover on Mars in summer 2021 (*Space Daily* 2021), but almost insignificant concern for navigating those activities with responsibility towards the burgeoning problem of space debris which is a threat to human survival (Thompson 2021).

Space/Atmospheric Drag and Greenhouse Effect

Discussed next is the factor that is not only linked with space debris accumulation but also directly linked with the greenhouse effect produced by Earth's pollution (NASA 2022). Greenhouse effect increases the temperature on Earth but in outer space they land up doing opposite. Hence, space drag is a phenomenon where increased carbon dioxide levels are cooling the upper atmosphere, which decreases the atmospheric density and results in debris particles stay longer in the orbit than they would normally have (Kaufman 2006). Space drag affects how long defunct satellites, spent rocket boosters, and other space debris stay in orbit, contributing to the space junk problem (Atkinson 2010). This raises the risk of collisions with satellites and makes it more hazardous to launch spacecraft. In other words, reduced Atmospheric Drag aggravates the problem of space debris.

As described in detail in the preceding sections, the number of satellites and space maneuvers in the LEO are increasing with time. Satellites, acted upon by various forces, orbit in space along a particular trajectory and the atmospheric drag is present as a force that plays an important role at various altitudes in space. Accounting for the atmospheric drag becomes even more important because underestimation or overestimation of the atmospheric drag could result in the incorrect modeling of the force that acts on the satellites (Gaposchkin and Coster 1998). The density plays an important role in the determination of the atmospheric drag (Majid, Owais, and Qureshi 2018). For example, the mass density of Earth's thermosphere between 90 and 600 kilometers altitude appears as a critical parameter for Low Earth Orbit (or LEO) prediction because of the atmospheric drag on satellites in this region (Emmert 2015). With increase in greenhouse emissions from earth, the resultant phenomenon of atmospheric drag is then worsening the problem of space debris very badly.

How Does the IPCC Look at Atmospheric Drag?

A key challenge before human race is that stabilizing greenhouse gas (GHG) concentrations will require large-scale transformations in human societies, from the way that we produce and consume energy to how we use the land surface and outer space. A natural question in this context is what will be that 'transformation pathway'? IPCC deliberations on Representative Concentration pathways shows what the optimum feasible transformative pathway could be. These are multiple possible scenarios presented in discussing different climate futures depending upon how well (or poorly) greenhouse emissions are controlled by human beings.

IPCC's Fifth Assessment Report (AR5) had categorized four Representative Concentration Pathways (RCPs) though RCP is a GHG concentration (not emissions) trajectory adopted by the IPCC. Four pathways were used for climate modeling and research for the IPCC fifth Assessment Report (AR5) in 2014 describe different climate futures, all of which are considered possible depending on the volume of GHGs emitted in the years to come. Representative Concentration pathways are characterized by the radiative forcing produced by the end of the 21st century. Radiative forcing is the extra heat the lower atmosphere will retain as a result of additional GHGs (Jubb, Canadell, and Dix 2022). The complexity of humanity's possible future emissions may be broadly extrapolated by a few representative pathways namely RCP1.9, RCP3.4, RCP4.5, RCP6.0, and RCP8.5.

As a general rule, lower indices of these RCPs are more sustainable for our environment. For example, RCP1.9 is a pathway that limits global warming to below 1.5°C and is the aspirational, but not feasible, goal of the Paris Agreement (Hausfather 2018). Similarly, RCP2.6 is a "very stringent" pathway. According to the IPCC, RCP2.6 requires that carbon dioxide (CO_2) emissions start declining by 2020 and go to zero by 2100. Feasibility is better represented by RCP3.4 that represents an intermediate pathway between the "very stringent" RCP2.6 and less stringent mitigation efforts associated with RCP4.5 (*Ibid.*). Most of the research done on RCPs agree that atmospheric concentrations in baseline scenarios for the year 2100 lie above the RCP6.0 (worryingly serious) but below the RCP8.5 (uncontrolled and disastrous). And human endeavor should aim to decrease GHG emission to RCPs with indices 3.4 or lower.

All these pathways are useful because they provide a limited number of baseline scenarios from which modeling can be performed and results across studies can be compared. It shows with clarity that the reduction in thermospheric density is predicted to continue at higher rates with further increases in drags. Hence, higher the RCP index, higher the greenhouse emission and higher the drag in outer space. This point is analyzed in better detail in the next section. Also, the impact that projected GHG emissions will have upon thermospheric neutral densities through to the year 2100 have been modeled with use of the *Whole Atmospheric Community Climate Model* with thermosphere and ionosphere extension (WACCM-X)

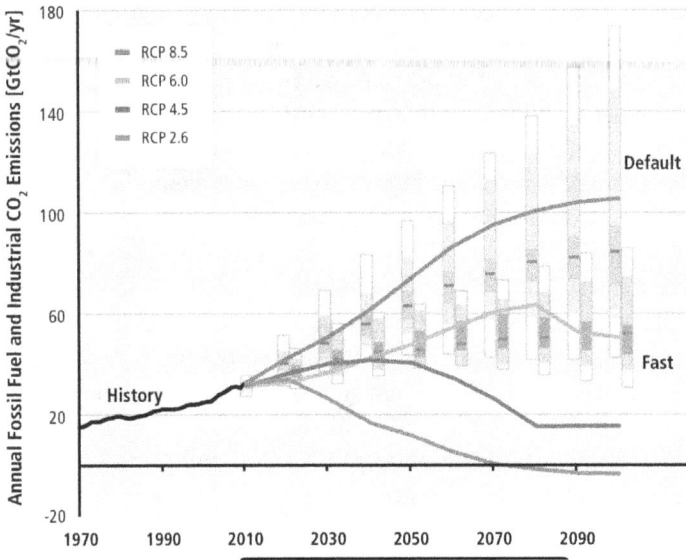

Figure 6: Greenhouse emissions under RCP 8.5.

Source: ipcc_wg3_ar5_chapter6.pdf.

(National Center for Atmospheric Research 2022). Orbital lifetimes for debris objects will increase threefold between the year 2000 and 2100 under RCP8.5 which arises from if no effort is made to reduce emissions and represents a failure to curb warming by 2100 (as illustrated in Figure 6). What it means is that if GHG mitigation efforts are not ramped up by a regime complex for climate change particularly for reducing greenhouse emissions, it would lead to a fatal threat to human survival. More greenhouse emissions on Earth will lead to more debris in outer space. This will have a significant impact on the debris environment.

The unified effort at reducing Greenhouse emission must be aimed at keeping the emissions below the predictions of pathways 6, most preferably below RCP4.5.

Sixth IPCC Assessment Report

The sixth IPCC Assessment Report released in August 2021 (Torok, Goldie, and Ashcroft 2021) was followed by its meetings on the second installment from the Working Group II's contribution to the Sixth

Assessment Report (IPCC 2022b). The report prepared by IPCC's Working Group II has further built on the Working Group I's contribution to the Sixth Assessment Report released in August 2021 that showed that climate change is widespread, rapid, and intensifying (IPCC 2022a). It shows that extreme weather is taking hold in every part of the planet, the atmosphere and seas are warming at rates unprecedented in human history and some of the consequences are irrevocable.

This definitely puts a challenge to yet another ominous warning found in the latest April 2022 IPCC report of the urgent need to slash emissions which otherwise set the world on track to reach 3.2°C by the end of the century (Sembiring 2022). All this is making experts debate in ever expanding frameworks of the plenary health prognosis. They believe that policymakers and government officials have to be among the first to see the merit of planetary health concept although civil society and corporations must also join in. By so doing, the overall impact on the health of the planet, which is currently measured by planetary boundaries, carbon budget, ecological footprint, among others, can be reduced, minimized, or even reversed (Sembiring 2022). All this debate has intensified focus on multifaceted assessments by the scientific community but perhaps all of this boiled down to a single critical message that the "time is running out" (Harvey 2021). Only drastic cuts in GHG emissions in this decade can prevent us from raising global temperatures to a disastrous levels of no-return. This is where cluttering of LEO becomes integral to climate change mitigation and adaptation strategies.

Conclusion

To conclude, therefore, apart from various endeavors by the IADC and the favorable RCP pathways endorsed by successive IPCC assessment reports, debates on building the Regime for Climate Change in Space have been strengthened by deploying the concepts like planetary health and Space 2.0 (Pyle 2019). These concepts seek to approach space through a transformational change in accessing and utilizing space with focus on hybrid and tactical satellites by a variety of partnerships of state actors and private players (Clendaniel 2014). Space 2.0 emphasizes on the "small, cheap and many" in terms of satellite design challenges the traditional approach of relying on a smaller number of large, expensive, and increasingly vulnerable satellites that must operate in an increasingly challenging and crowded space domain (Davis 2018). It also emphasizes on building cost effective and responsive space launch capabilities, best epitomized by Elon Musk's SpaceX that now regularly launches reusable rockets (Davis 2018). The cost effectiveness, hybrid use, and reusability can tackle the cluttering of space effectively.

Space technologies have no doubt improved human's ability to detect the impact of human activity on climate change. But the Space 2.0 debates suggest to measure and rectify the negative impact of climate change using two strategies namely, (i) Augmentation (ii) Reconstitution (Davis 2020, 2022). Here, augmentation refers to developing the capabilities of an existing space mission instead of launching new ones. In turn, augmentation may help counter the space race dilemma between countries as the focus on achieving space capabilities may be acquired not by sending more and more satellites and rockets but by augmenting the capabilities of those already in use. Most importantly, by sending fewer space rockets and satellites, an effort can be made to reduce the ever increasing debris problem as well. The second technique is called Reconstitution and refers to reducing the number of space satellites and rockets by reconstituting and reusing the ones already present. Both these techniques taken together are likely to take on the issue of rising space debris by cluttering the LEO as well as by the result of greenhouse emissions that trigger space drag and thereby exacerbate the problem of space debris.

Second, the use of Augmentation and Reconstitution will bring about a cut in expensive technologies. Hence, another characteristic of Space 2.0 focuses on sharing the technology with partners. Space capabilities can be developed and launched from sites in ways that meet both the operational requirements of both the host country and its coalition partners. As costs fall, small satellites costing a few million dollars will be able to do the job that once required large, complex, multibillion-dollar satellites. Space 2.0 looks at space in terms of a convergence matrix rather than a field for high action race for maximization of capabilities. It does sound like an idealistic view of the outer space, but indeed presents a strong norm to develop and abide by. These two strategies can help reduce the inevitable space security dilemma and thereby reduce space debris. Along with other innovations including artificial intelligence (AI), 5G, the Internet of Things (IoT) and robotics — Space 2.0 offers further potential for supporting the fight against climate change in outer space. Collectively such technologies could play a fundamental role in meeting the Paris Agreement's target of limiting global warming to 1.5°C (under the RCP 2.6) and reducing the risk of a potentially dangerous space collision between satellites and millions of debris particles orbiting the Earth. The study recommends that the concept of Space 2.0 may be introduced in the international regime complex for climate change for space backed by strong space policy and space norms be created to make the international community abide by the same.

References

Anz-Meador, P. (2019). *Root Cause Classification of Breakup Events 1961–2018.* https://ntrs.nasa.gov/citations/20190033947.

Atkinson, N. (2010). Climate change contributes to space junk problem. *Universe Today.* https://www.universetoday.com/67246/climate-change-contributes-to-space-junk-problem#:~:text=This%20in%20turn%20affects%20how%20long%20defunct%20satellites%2C,to%20drop%20out%20of%20orbit%20and%20burn%20up.

Brahmand (2009). Space station safe from cosmos 1275 debris: NASA. *Brahmand News.* https://brahmand.com/news/Space-station-safe-from-Cosmos-1275-debris-NASA/1369/1/23.html.

Bruno, R. and Perez, D. (2017). 7th European Conference on space debris. https://conference.sdo.esoc.esa.int/proceedings/sdc7/paper/341/SDC7-paper341.pdf.

Clendaniel, E. (2014). Clendaniel: Space 2.0. Where will the next technological wave take us? https://www.mercurynews.com/2021/11/06/clendaniel-the-next-phase-of-space-technology-advancement-is-upon-us/.

Davis, M. (2018). Space 2.0 — Why it matters for Australia's defence. *The Strategist,* https://www.aspistrategist.org.au/space-2-0-matters-australias-defence/.

Davis, M. (2022). Could space systems improve Australia's long-range strike capabilities? *The National Interest.* https://nationalinterest.org/blog/reboot/could-space-systems-improve-australia%E2%80%99s-long-range-strike-capabilities-199836.

Day, D. (2013). *The Space Review: China's ASAT enigma.* https://www.thespacereview.com/article/2251/1.

Emmert, J. T. (2015). Thermospheric mass density: A review. *Advances in Space Research, 56,* 773–824.

Gaposchkin, E. M. and Coster, A. J. (1988). Analysis of satellite drag. *The Lincoln Laboratory Journal. 1*(2), 203–224.

Gheorghe, A. V. and Yuchnovicz, D. E. (2015). *The Space Infrastructure Vulnerability Cadastre: Orbital Debris Critical Loads.* New York: SpringerLink. https://link.springer.com/article/10.1007/s13753-015-0073-2.

Gong, Z. (2016). https://www.unoosa.org/documents/pdf/copuos/stsc/2016/tech-21E.pdf.

Gorman, A. (2022). NASA says 27,000 pieces of space junk; first was in 1958. What is being done now? *Hindustan Times.* https://tech.hindustantimes.com/

tech/news/nasa-says-27-000-pieces-of-space-junk-first-was-in-1958-what-is-being-done-now-71643702913413.html.

Harris, G. (2014). Space debris expert warns of increasing CubeSat collision risk. https://phys.org/news/2014-09-space-debris-expert-cubesat-collision.html.

Harvey, F. (2021). What is the IPCC and why is its new climate report different from others? *The Guardian.* https://www.theguardian.com/environment/2021/aug/09/what-is-ipcc-why-new-climate-report-different.

Hausfather, Z. (2018). Explainer: How "Shared Socioeconomic Pathways" explore future climate change — Carbon brief. https://www.carbonbrief.org/explainer-how-shared-socioeconomic-pathways-explore-future-climate-change/.

IADC (2017). Statement on large constellations. https://iadc-home.org/what_iadc, accessed on May 15, 2022.

IADC (2019). https://iadc-home.org/what_iadc, accessed on May 15, 2022.

IADC (2021). https://www.iadc-home.org/documents_public/file_down/id/5249, accessed on May 15, 2022.

IPCC (2022a). Fifty-Fifth Session of the IPCC. Doc.-4-Rev.1-Approved-Summary-for-Policymarkers.pdf. https://www.ipcc.ch/site/assets/uploads/2022/03/Doc.-4-Rev.1-Approved-Summary-for-Policymarkers.pdf.

IPCC (2022b). https://www.ipcc.ch/report/ar6/wg2/downloads/report/. IPCC_AR6_WGII_FinalDraft_FullReport.pdf.

Jubb, I., Canadell, P., and Dix, M. (2022). Representative Concentration Pathways (RCPs). https://www.cawcr.gov.au/projects/Climatechange/wp-content/uploads/2016/11/ACCSP_RCP.pdf.

Kaufman, M. (2006). Greenhouse effect could cause space problem. *The Seattle Times.* https://www.seattletimes.com/nation-world/greenhouse-effect-could-cause-space-problem/.

Krasner, S. D. (1991). Global communications and national power: Life on the Pareto Frontier *World Politics*, *43*(3), 336–366.

Liao, X. L.W. (2016). *The Regime Complex of Global Space Governance.* Ghent: Ghent Institute for International Studies. https://biblio.ugent.be/publication/8501226/file/8501234.pdf.

Liberatore, S. and Chadwick, J. (2021). 21-ton Chinese rocket is tumbling to Earth and could shower debris on populated areas. *Daily Mail Online.* https://www.dailymail.co.uk/sciencetech/article-9537979/A-21-TON-Chinese-rocket-tumbling-Earth-shower-debris-populated-areas.html.

Majid, M., Owais, N., and Qureshi, N. (2018). Aerodynamic drag computation of Lower Earth Orbit (LEO) satellite. *Journal of Space Technology.* 12.-aerodynamic-drag-computation-of-lower-earth-orbit-(leo)-satellites.pdf.

McDowell, J. (2020). The globalization of space. Global16.pdf, https://planet4589.org/talks/global/global16.pdf.

McDowell, J. (2021). Uncontrolled reentry: Why China "just not caring" is a huge problem for space. https://www.inverse.com/science/long-march-5b-uncontrolled-reentry.

McDowell, J. (2022). *Jonathan's Space Report. GCAT.* https://planet4589.org/space/gcat/.

McDowell, J. C. (2020). General catalog of artificial space objects, Release 1.2.1, accessed on February 18, 2022. https://planet4589.org/space/gcat.

NASA (2012). Chronology of defining events in NASA history. https://www.history.nasa.gov/40thann/define.htm.

NASA (2015). Vanguard Satellite, 1958. https://www.nasa.gov/content/vanguard-satellite-1958.

NASA (2021). Space Debris and Human Spacecraft. https://www.nasa.gov/mission_pages/station/news/orbital_debris.html.

NASA (2022). What is the greenhouse effect? — Climate change: Vital signs of the planet. https://climate.nasa.gov/faq/19/what-is-the-greenhouse-effect/.

NASA Headquarters Library, accessed on January 19, 2022.

National Center for Atmospheric Research (2022). WACCM-X, High Altitude Observatory. https://www2.hao.ucar.edu/modeling/waccm-x.

Orbital Debris Program Office, NASA, accessed on January 12, 2022.

Pappas, S. (2018). What will happen to China's Tiangong-1 as it falls through the atmosphere? *Space.* https://www.space.com/40096-tiangong-space-station-destroyed-reentry-atmosphere.html.

Pyle, R. (2019). Space 2.0 pyle_space-2-0_press-kit.pdf. https://cdn.ces.tech/ces/media/pdfs/garys-bookclub-media/pyle_space_2-0_press-kit.pdf.

Roy, B. (2010). Contexts of interdisciplinary: Interdisciplinarity and climate change. In Roy, B. *et al.*, Interdisciplinarity and Climate Change: Transforming Knowledge and practice for Our Global Future. London: Routledge.

Sembiring, M. (2022). Planetary health: Managing competing tensions. *RSIS Commentary*, No.052/2022.

Solomon, S. C. *et al.* Anomalously low solar extreme-ultraviolet irradiance and thermospheric density during solar minimum. *Geophysical Research Letters*, 37, L16103.

Space Daily (2021). China's space achievements out of this world. https://www.spacedaily.com/reports/Chinas_space_achievements_out_of_this_world_999.html.

Tellis, A. J. (2019). India's ASAT test: An incomplete success — Carnegie endowment for international peace. https://carnegieendowment.org/2019/04/15/india-s-asat-test-incomplete-success-pub-78884.

Thompson, A. (2021). As China makes space strides, debris problem gains urgency. *Space News, Al Jazeera.* https://www.aljazeera.com/economy/2021/5/17/as-china-makes-space-strides-debris-problem-gains-urgency.

Torok, S., Goldies, J., and Ashcroft, L. (2021). Communicating climate change has never been so important, and this IPCC report pulls no punches. https://phys.org/news/2021-08-climate-important-ipcc.html.

US Department of State (2021). U.S.–China Joint Glasgow Declaration on enhancing climate action in the 2020s, accessed om November 10, 2021. https://www.state.gov/u-s-china-joint-glasgow-declaration-on-enhancing-climate-action-in-the-2020s/.

Section II

Institutions and Initiatives

https://doi.org/10.1142/9789811263750_0007

Chapter 7

Climate Action by the European Union: Making the European Green Deal a Reality

Kakoli Sengupta*

Abstract

The European Union (EU) has tasked itself to make Europe the first climate neutral continent in the world. In December 2019, the European Commission presented the European Green Deal committing to climate neutrality by 2050, with the objective of transforming itself from a high to a low-carbon economy, without reducing prosperity while simultaneously improving people's quality of life, through better health and cleaner air and water. The EU however also believes that climate change must be addressed by a global response and thus, actively engages and supports its international partners in climate action, in particular through the UN Framework Convention on Climate Change (UNFCCC) and the Paris Agreement. This chapter examines various relevant elements and legislations encompassed within the European Green Deal aimed at guiding and ensuring this transformational change. It contends that as a global climate leader, the EU must also make every effort to reassure countries in the Global South that its efforts at this ambitious green

---

* Kakoli Sengupta is Associate Professor and Former Head of Department, Department of International Relations, Jadavpur University, Kolkata, West Bengal, India.

transition will be inclusive. A strong framework for doing this is through the Global Gateway, which was already announced in late 2021. If the Global Gateway project succeeds, this will be a major climate changer for the world.

Keywords: European Union, Green Deal, Legislations, Climate Change, Global Gateway

Introduction

Climate Action has come to be at the centerstage of the global politics. The norm-driven European Union (EU) in particular recognizes the grim reality that climate change and environmental degradation pose an existential threat not just to Europe but to the entire world. Articles 191–193 of the Treaty on the Functioning of the European Union (TFEU) confirm as also specify EU competencies in the area of climate change. In accordance with Articles 191 and 192(1) of the TFEU, the EU is stipulated to contribute to the pursuit of the following global objectives: preserving, protecting, and improving the quality of the environment, promoting measures at international level to deal with regional or worldwide environmental problems, and in particular combating climate change (European Union 2020a).

The expressed goal of the EU has been to make Europe the first climate neutral continent in the world. In this, the European Green Deal outlines the blueprint for this transformational change (European Union 2019a). In December 2019, the European Commission had presented the European Green Deal committing to climate neutrality by 2050. On July 14, 2021, the European Commission adopted a set of proposals to make the EU's climate, energy, transport, and taxation policies fit for reducing net greenhouse gas emissions by at least 55% by 2030, compared to 1990 levels. Achieving these emission reductions in the next decade is crucial to Europe for becoming the world's first climate-neutral continent by 2050 and thus making the European Green Deal a reality (European Union 2019b).

European Green Deal

The main aim of the European Green Deal is to transform the EU from a high to a low-carbon economy, without reducing prosperity while simultaneously improving people's quality of life, through better health and cleaner air and water (*The Guardian* 2020). The EU believes that emissions must be reduced in all sectors, from industry and energy, to transport and farming which will help in achieving its ambitious decarbonization objectives. Climate change, of course, must be addressed by a global response as it is a global threat and, therefore, the EU also actively engages and supports its international partners in climate action, in particular through the UN Framework Convention on Climate Change (UNFCCC) and the Paris Agreement (European Union 2020b). The main elements of the EU Green Deal include: climate action, clean energy, sustainable industry, buildings and renovations, sustainable mobility, eliminating pollution, farm to fork, preserving biodiversity, research and development. Preventing unfair competition from carbon leakage (European Union 2021).

The European Climate Law requires that all EU policies contribute to achieving the EU Green Deal objective. The EU Commission is thus reviewing every EU law to ensure its alignment with the EU emission reduction targets, under an exercise termed the "Fit for 55 package" (European Union 2021). Some of the key legislations that the EU Commission proposes to revise in light of the revised emissions reduction target include: the Renewable Energy Directive, the Energy Efficiency Directive, the Emissions Trading System, the Effort Sharing Regulation, the Land Use, Land Use Change and Forestry Regulation, the Energy Performance of Buildings Directive, and the Energy Taxation Directive (European Union 2021). Second, EU's Clean Energy policy aims to develop a power sector based largely on renewable sources. The Offshore Renewable Energy Strategy encourages the investment of almost €800 billion within 2050 in offshore energy infrastructure and research. The Clean Energy for all Europeans package will facilitate the strategy for energy system integration, which aims to improve the coordination of planning and operation of the energy system "as a whole", across multiple energy carriers, infrastructure, and end uses (European Union 2020a).

Trans-European Energy Networks

The European Commission has also proposed a revision of the Regulation on the Trans-European Energy Networks (TEN-E). The TEN-E Regulation established a new approach to cross-border energy infrastructure planning when it was adopted in 2013. It has brought together multiple stakeholders in regional groups to select and help implement projects of common interest (PCIs) that link member states' energy networks, connect regions currently isolated from European energy markets, strengthen existing cross-border interconnections, and help integrate renewable energy (European Union 2020b). The proposed revision by the Commission aims at modernizing and upgrading the TEN-E framework, reflecting the Green Deal objectives and making it fit for the infrastructure needs of the clean energy system of the future. The Regulation will now focus on the 2050 climate neutrality objective under the European Green Deal, and to adapt to the rapid ongoing technological developments.

The proposed changes in EU's TEN-E framework reflect the key role for its energy infrastructure that it is expected to play in this green transition. New and updated infrastructure categories and a new approach to infrastructure planning will support the role of electrification in the future energy mix, help to decarbonize the gas sector through renewable and low-carbon gases, including hydrogen, and develop a more integrated energy system (European Union 2021). The EU had already adopted a new Circular Economy Action Plan in March 2020. It is one of the main building blocks of the European Green Deal, Europe's new agenda for sustainable growth. The EU's transition to a circular economy will reduce pressure on natural resources and will create sustainable growth and jobs. The new action plan announces initiatives along the entire life cycle of products. It targets how products are designed, promotes circular economy processes, encourages sustainable consumption, and aims to ensure that waste is prevented and the resources used are kept in the EU economy for as long as possible. The Circular Economy Action Plan aims to:

- make sustainable products the norm in the EU;
- empower consumers and public buyers;
- focus on the sectors that use most resources and where the potential for circularity is high such as: electronics and ICT, batteries and vehicles, packaging, plastics, textiles, construction and buildings, food, water and nutrients;
- ensure less waste;
- make circularity work for people, regions and cities; and
- lead global efforts on circular economy (European Union 2021b).

The Trans-European Energy Networks are slated to work closely with EU's Circular Economy Action Plan that includes a Sustainable Products Policy that regulates the improvement of product reusability, reparability and integration of recycled contents (European Commission 2019). On March 10, 2020, the Commission had laid the foundations for an industrial strategy that supports the twin transition to a green and digital economy, making EU's industry more competitive globally, and enhance Europe's open strategic autonomy (European Union 2021b). The aim of the EU Industrial Strategy is to develop markets for climate neutral and circular products and to encourage the digital transition inside the EU. This is where EU's Green Deal denotes these measures that are necessary to ensure the supply of the critical raw materials needed for clean technologies such as clean hydrogen, fuel cells, and other alternative fuels, energy storage, and carbon capture, storage, and utilization (European Commission 2019).

Transitioning to Energy Efficient Buildings

Buildings have been identified as the largest sources of energy consumption in Europe and responsible for over a third of EU emissions. However, only 1% of buildings across Europe have undergone energy-efficient renovation, hence effective action is crucial to making Europe climate-neutral, i.e., net zero emissions by 2050. Therefore, renovating both public and private buildings has been outlined in the Green Deal as a key initiative to drive energy efficiency in the sector. To pursue this dual ambition of energy gains and economic growth, in 2020 the Commission had published its new strategy to boost renovation titled *A Renovation Wave for Europe — Greening Our Buildings, Creating Jobs, Improving Lives*. Doubling annual energy renovation rates in the next 10 years is the aim of this strategy. Apart from reduction of emissions, these renovations will enhance the quality of life living in and using the buildings and should create many additional green jobs in the construction sector. Since nearly 34 million people are unable to heat their homes properly, renovation also addresses the issue of energy poverty. The health and well-being of vulnerable people is also addressed while reducing their energy bills. This is also part of the renovation wave strategy. The renovation wave initiative will build on measures agreed under the "Clean Energy for All Europeans" package, notably the requirement for each EU country to publish a long-term building renovation strategy, other aspects of the Directive on Energy Performance of Buildings, and building-related aspects of each EU country's national energy and climate plans (Europa 2019).

This plan of renovation of buildings to make their energy efficient in European Green Deal is expected to create new opportunities for innovation and investment and jobs as well as reduce emissions, create jobs and growth, address energy poverty, reduce external energy dependency, and improve overall health and well-being of European citizens. At the same time, it will ensure opportunities for all, support vulnerable citizens by tackling inequality, and strengthen the competitiveness of European companies. The European Commission through the Green Deal strives to transition toward greener mobility which will offer clean, accessible and affordable transport even in the most remote areas (European Union 2021b). Second, the Sustainable Mobility policy area also comprises initiatives to reduce transport emissions, which account for 25% of the EU's greenhouse gas emissions. The Strategy for Sustainable and Smart Mobility lays the foundation for action to transform the EU transport sector, with the aim of a 90% cut in emissions by 2050, delivered by a smart, competitive, safe, accessible, and affordable transport system. Increased capacity and decreased congestion and pollution could all be attained as a result of efforts to promote more sustainable means of transport. The strategy sets a number of targets to 2030 including:

- at least 30 million zero-emission cars will be in operation on European roads;
- 100 European cities will be climate neutral;
- high-speed rail traffic will double across Europe;
- scheduled collective travel for journeys under 500 km should be carbon neutral;
- automated mobility will be deployed at large scale; and
- zero-emission marine vessels will be market-ready, with further targets to 2035 and 2040 (European Union 2021b).

Renovating Europe's Transport Network

The EU Commission has also planned a revision of the Regulation on the trans-European transport network (the TEN-T Regulation) and of the Directive on intelligent transport systems. This aims to increase the uptake of zero-emission vehicles, make sustainable alternative solutions, and support digitalization and automation (Europa 2020). Besides, the Commission has proposed that from 2026, road transport will be covered by emissions trading which will stimulate cleaner fuel use and re-invest in clean technologies. The Commission has also proposed carbon pricing for the aviation sector which had until recently benefited from an exception. It has also proposed to promote sustainable aviation fuels and there is an obligation for planes to take on sustainable blended fuels for all departures from EU airports. The European Commission has also proposed to extend carbon pricing to the maritime sector. Targets will be set by the Commission for major ports to serve vessels with onshore power thereby reducing the use of polluting fuels for which local air quality is harmed (Europa 2019). Also, European industry has been presented with a unique opportunity as markets will be created for clean technologies and products. Energy, transport, construction, and renovation sectors will see a massive transformation because of the green transition. Around 35 million buildings could be renovated by 2030. Around 1,60,000 additional green jobs could be created in the construction sector by 2030 (European Union 2021b).

The largest environmental cause of multiple mental and physical diseases is pollution. It is also a significant driver of biodiversity loss. The EU Commission has proposed a *Zero Pollution Action Plan* which proposes that pollution elimination measures are incorporated into all policy developments and steps are taken to further decouple economic growth from the increase of pollution. The action plan comprises a Chemical strategy for sustainability to protect the environment against hazardous chemicals, a Zero pollution action plan for water, air and soil, to better prevent, remedy, monitor, and report on pollution and the revision of measures to address pollution from large industrial installations to ensure that they are consistent with related EU Green Deal objectives (European Union 2021b). At the heart of the European Green Deal is the Farm to Fork strategy which aims to make food systems fair, healthy, and environmentally friendly. The Farm to Fork Strategy aims to accelerate the transition to a sustainable food system that should:

- have a neutral or positive environmental impact;
- help to mitigate climate change and adapt to its impacts;
- reverse the loss of biodiversity;
- ensure food security, nutrition and public health, making sure that everyone has access to sufficient, safe, nutritious, sustainable food; and
- preserve affordability of food while generating fairer economic returns, fostering competitiveness of the EU supply sector, promoting fair trade (European Union 2021b).

Agriculture: Farm to Fork Strategy

EU's Farm to Fork strategy focuses on reducing waste as also transforming various components of the manufacturing, processing, retailing, packaging, and transportation of food. The strategy proposes to spend €10 billion on research and innovation on food, bio-economy, natural resources, agriculture, fisheries, aquaculture, and the environment, as well as digital technologies and nature-based solutions for agri-food, funded by Horizon Europe, the EU's research and innovation framework program (European Union 2021b). Second, the EU Biodiversity Strategy for 2030 identifies changes in land and sea use, overexploitation, climate change, pollution, etc., as the principal drivers in biodiversity loss. The European Commission identifies construction, agriculture and food and drink sectors as the industries highly dependent on biodiversity. The EU Biodiversity strategy will work together with the Farm to Fork strategy by focusing on restoring forests, soils, and wetlands and creating green spaces in cities (European Union 2021b).

Research and Development remains integral to the EU's Green Deal. New technologies are promoted to harness several of the Green Deal initiatives. Many research and development initiatives will be funded by Horizon Europe. Under Horizon Europe, the European Union will form green partnerships with various industries and its member states to focus on key areas such as low-carbon steel, batteries, clean hydrogen, environment, and biodiversity (European Union 2021b). All this involves a whole range of initiatives to achieve a significant reorientation in EU's economy toward a low carbon model due to the EU Green Deal. This, of course, entails the risk of continued carbon leakage. For this, the EU Commission has identified risks either in their production which is being transferred from Europe to other countries with lower ambition for emission reduction, or that EU products are replaced by more carbon-intensive imports (European Union 2021b). Such carbon leakage can shift emissions outside of Europe and therefore seriously undermine EU and global climate efforts (European Union 2021b).

The EU Commission has proposed a Carbon Border Adjustment Mechanism to ensure that the price of imports reflects more accurately their carbon content (European Union 2021b). It is a climate measure that should prevent the risk of carbon leakage and support the EU's increased ambition on climate mitigation, while ensuring WTO compatibility. The Carbon Border Adjustment Mechanism will equalize the price of carbon between domestic products and imports and ensure that the EU's climate objectives are not undermined by production relocating to countries with less ambitious policies (Europa 2021a). The mechanism has been designed in compliance with World Trade Organization (WTO) rules and other international obligations of the EU. EU importers will buy carbon certificates corresponding to the carbon price that would have been paid, had the goods been produced under the EU's carbon pricing rules. Conversely, once a non-EU producer can show that they have already paid a price for the carbon used in the production of the imported goods in a third country, the corresponding cost can be fully deducted for the EU importer. Thus, in this way, the Carbon Border Adjustment Mechanism will help reduce the risk of carbon leakage by encouraging producers in non-EU countries to green their production processes (Weforum 2021).

Over €1 trillion of investments are projected to be required to finance all these initiatives of the EU Green Deal. As envisaged by the European Commission, half of this finance is coming from the EU Budget and the EU Emission Trading Scheme and half from Invest EU initiative. To meet the objective of carbon-neutrality by 2050, the European Commission recently put forward its new *Sustainable Finance Strategy* to channel private financial flows into relevant sustainable economic activities (Weforum 2021). The aim of the new sustainable finance strategy aims to support the financing of the transition to a sustainable economy by proposing action in four number of areas: transition finance, inclusiveness, resilience and contribution of the financial system, and global ambition (*The Third Pole* 2021). All this is explained through the European Green Deal that showcases how Europe has this unique opportunity to create sustainable and inclusive growth for present and future generations to come. This is where EU's Green Deal represents innovation, technology, growth, and hope for Europe.

Europe and the Glasgow COP26

As was seen at the Glasgow COP, the biggest challenge for EU's Green Deal comes from carrying forward UNFCCC which calls for political will and cohesion among all of the European nations especially to see that the proposals of the Deal, which are aimed at saving the environment and ensuring that a "green Europe" goes hand in hand with technology are implemented as planned (Chatham House 2021). The Paris Agreement aims to keep the rise in the global average temperature to "well below" 2° above pre-industrial levels, ideally 1.5°; strengthen the ability to adapt to climate change and build resilience; and align all finance flows with "a pathway towards low greenhouse gas emissions and climate-resilient development". According to the Paris Agreement, countries themselves decide by how much they will reduce their emissions by a certain year. They communicate these targets to the UNFCCC in the form of "nationally determined contributions", or "NDCs". One of the main yardsticks for success at the COP26 Conference held at Glasgow is that as many governments as possible submitted new NDCs and, when put together, these are ambitious enough to put the world on track for "well below" 2°, preferably 1.5°. The U.K.'s overarching aim for the Glasgow summit is to "keep 1.5 degrees alive" (Chatham House 2021).

Success of COP26 is of immense importance to the European Union. As stated by the European Commission President, Ursula von der Leyen, "The European Union will bring to Glasgow the highest level of ambition. We do it for all future generations. We do it for our planet. And we do it for Europe" (Europa 2021b). With the European Green Deal, the EU is setting a positive example at the prestigious and extremely important summit which will be attended by the President of the European Commission and members of the College of Commissioners who will be present at the Summit along with other world leaders (Open Democracy 2021). However, the European Green Deal has been subjected to criticism which generally fall into three categories — the first is that the European Deal clings to the idea that absolute decoupling of economic growth and environmental impact is possible. That is, it is assumed that "green" economic growth can be achieved at the same time as dramatically reducing CO_2 emissions. By outsourcing polluting activities, absolute decoupling can be made to work on a regional level but there is no evidence that this can happen on a global scale. The second criticism is the plan's blind faith in technology.

Renewables without doubt promise to provide a windfall in addressing reductions in fossil fuels though they can also create multiple dependencies on scarce raw materials. For instance, this could lead to a shortage of lithium, cobalt, nickel, and other rare earth metals by 2050 — elements that are mainly concentrated in the Democratic Republic of Congo, Argentina, Chile, Bolivia, Indonesia, the Philippines, Australia, and China. The final criticism is that the financial frameworks promoted by the European Green Deal subordinate the public interest to private gain.

The availability of public funds has prompted the large European energy companies like BP, Shell, Total, and Repsol to accelerate their greenwashing apparatus to present themselves as essential actors in the recovery and transition. The structure of the funds themselves effectively squeezes out other actors such as small and medium-sized enterprises. It has been criticized that the timelines are too fast, the funding volumes too large, and the process too bureaucratic (Stanly 2021).

Challenges of Renewable Resources

In recent times, one worrying issue besetting Europe has been the soaring natural gas prices. Many European countries have moved away from coal to gas to produce electricity due to Europe's shift toward cleaner energy. This has increased Europe's reliance on gas. However, over the years, natural gas production has shrunk in Europe as many countries have shut down production fields over environmental concerns. Europe's main producer of gas, Norway, has seen its production shrink from 117.6 bcm to 105.3 bcm in 2021. This has made Europe largely dependent on Russia (Stanly 2021). Ukraine war of 2022 was to additionally reveal political costs of such dependencies. Due to the growing global demand and falling production in Europe, prices have risen in general. In the meanwhile, the situation has been worsened due to the fact that supplies from Russia via a pipeline that runs through Ukraine and Poland also shrank. Russia has also built another gas pipeline, Nord Stream 2, which will take Russian gas directly to Germany bypassing Ukraine and Poland, whose governments are critical of the Kremlin. However, European authorities are yet to give approval to the Nord Stream 2 pipeline. Several European countries as well as the United States have been critical of the pipeline as this, in their opinion, will enhance Russia's leverage over Europe and would also permit Russia to economically punish Poland and Ukraine (Stanley 2021).

The year 2022 saw Gazprom, the Russia's state-controlled energy company that supplies about 35% of Europe's gas requirements, booking less additional exports than wanted by traders. Trends like these add pressure on supplies. Speculation has been fueled that Russian President Vladimir Putin is trying to use the energy crisis in Europe to get approval for the Nord Stream 2 pipeline from the EU (Stanly 2021). However, Russia has dismissed this criticism saying that it had no role in the energy crisis. Markets were temporarily calmed by the Russian leader Mr. Putin's statement that Russia could "reach another record of deliveries of our energy resources to Europe, including gas" (*Euronews* 2021). Hungarian Prime Minister Victor Orban recently blamed European Union action to combat climate change for a rise in energy prices. The steep jump in energy prices has stoked tensions between the EU nations. Wealthy EU countries believe it proves the need to press on with new climate change policies while poorer EU countries are concerned that they could push up consumer bills.

Europe's Evolving Demand and Supply

Victor Orban, pro-Russia long-time prime minister of Hungary told state radio that "bureaucrats in Brussels" are fighting against climate change by continuously raising the price of energy generated using coal and gas, which has led to a surge in households' energy bills across Europe (*Euronews* 2021). Orban stated, "These (EU) decisions must be withdrawn ... at present gas prices are where they should be in 2035. Brussels is not the solution today, they are the problem" (*Reuters* 2021). He even dismissed such plans as "utopian fantasy" (*Euronews* 2021). He blamed the EU Climate Policy Chief Frans Timmermans for the crisis and stated, "Its a Commissioner called Timmermans who is posing the biggest threat to us" (*Euronews* 2021). Frans Timmermans however has said that emissions from Europe's transport sector are on a rising trend and if they were left unchecked, that would prevent the EU goal to cut net greenhouse gas emissions by at least 55% by 2030 (*Reuters* 2021). Orban's stance is at odds with other EU countries who say the price jump should trigger a faster switch to low-emission, locally produced renewable energy, to help reduce exposure to imported fossil fuel prices (Reuters 2021).

On October 13, 2021, the European Commission adopted a *Communication on Energy Prices*, to tackle the exceptional rise in global energy prices, which is projected to last through the winter, and help Europe's people and businesses. The Communication included a "toolbox" that the EU and its member states can use to address the immediate impact of current prices increases, and further strengthen resilience against future shocks. Energy Commissioner Kadri Simson while presenting the toolbox said: "Rising global energy prices are a serious concern for the EU. As we emerge from the pandemic and begin our economic recovery, it is important to protect vulnerable consumers and support European companies. The Commission is helping Member States to take immediate measures to reduce the impact on households and businesses this winter. At the same time, we identify other medium-term measures to ensure that our energy system is more resilient and more flexible to withstand any future volatility throughout the transition.

The current situation is exceptional, and the internal energy market has served us well for the past 20 years. But we need to be sure that it continues to do so in the future, delivering on the European Green Deal, boosting our energy independence and meeting our climate goals" (European Commission 2021a). Among the *immediate measures* to protect consumers and businesses, the Commission has listed the following:

- Provide emergency income support for energy-poor consumers, for example through vouchers or partial bill payments, which can be supported with EU ETS revenues.
- Authorize temporary deferrals of bill payments.
- Put in place safeguards to avoid disconnections from the grid.
- Provide temporary, targeted reductions in taxation rates for vulnerable households.
- Provide aid to companies or industries, in line with EU state aid rules.
- Enhance international energy outreach to ensure the transparency, liquidity, and flexibility of international markets.
- Investigate possible anti-competitive behavior in the energy market and ask the European Securities and Markets Authority (ESMA) to further enhance monitoring of developments in the carbon market.
- Facilitate a wider access to renewable power purchase agreements and support them via flanking measures.

Among the *medium-term measures* listed by the Commission for a decarbonized and resilient energy system were:

- Step up investments in renewables, renovations and energy efficiency and speed up renewables auctions and permitting processes.
- Develop energy storage capacity, to support the evolving renewables share, including batteries and hydrogen.

- Ask European energy regulators (ACER) to study the benefits and drawbacks of the existing electricity market design and propose recommendations to the Commission where relevant.
- Consider revising the security of supply regulation to ensure a better use and functioning of gas storage in Europe.
- Explore the potential benefits of voluntary joint procurement by Member States of gas stocks.
- Set up new cross-border regional gas risk groups to analyze risks and advise Member States on the design of their national preventive and emergency action plans.
- Boost the role of consumers in the energy market, by empowering them to choose and change suppliers, generate their own electricity, and join energy communities (European Commission 2021a).

Challenges in Energy Transitioning

According to the Commission, the clean energy transition is the best insurance against price shocks in the future, and needs to be accelerated. The EU will continue to develop an efficient energy system with high share of renewable energy. The Commission believed that "the measures set out in the toolbox would help to provide a timely response to the current energy price spikes, which are the consequence of an exceptional global situation. They will also contribute to an affordable, just and sustainable energy transition for Europe, and greater energy independence" (*The Times of India* 2021). On October 20, 2021 the European Commission President Ursula von der Leyen urged the 27 member bloc to wean themselves off natural gas to make the bloc a more independent player in the world and to speed up the transition to clean energy. She told legislators that the EU is made more vulnerable by the fact that it imports more than 90% of its gas of which a major part comes from strategic rival Russia. As a result, she wanted the EU to double down on a swift transition to clean energy from wind and sun which can be produced domestically and also, in the long run, would be ultimately a lot cheaper than imported fossil fuels. She said, "the transition to clean energy is not only vital for our planet. It is also crucial for our economy and for the resilience to energy price shocks" (*APN News* 2021).

On October 25, 2021 nine European Union countries including Germany released a *statement* stating that they will not support an overhaul of the electricity market. The written statement stated, "As the price spikes have global drivers, we should be very careful before interfering in the design of internal energy markets". It also stated, "This will not be a remedy to mitigate the current rising energy prices linked to fossil fuels markets". The statement released by Luxembourg, Austria, Germany, Denmark, Estonia, Finland, Ireland, Latvia, and the Netherlands said, "We need a well-integrated EU energy market that functions based on market mechanisms and good interconnections as part of the solution to strengthen the resilience to price shocks" (European Council on Foreign Relations 2021a).

Analysts believe that that the green transition presents a series of opportunities and risks for Europe. The transition to renewable energy and other efforts to eliminate carbon emissions are likely to increase the European Union's dependence on networked technologies and raw materials. The EU needs a strategy to manage this change in a way that maintains the fragile consensus on the European Green Deal between EU member states, and that fulfills its ambitions for global climate leadership. As a global climate leader, the EU should make every effort to reassure countries in the global south that the green transition will not leave them behind. A strong framework for doing this is through the Global Gateway (European Council on Foreign Relations 2021b). On September 15, 2021, the European Commission had announced that the European Union would implement a connectivity grand strategy called the *Global Gateway*. The strategy is designed to counter China's $1 trillion Belt and Road Initiative, which is reshaping the architecture of global commerce. Commission President Ursula von der Leyen while introducing the Global Gateway emphasized partnerships with African countries and identified the February 2022 EU–Africa Summit as the first venue in which the EU will discuss its new connectivity strategy with regional partners.

Global Gateway: The Way Ahead

According to analysts, the key to ensuring that Europe becomes Africa's leading growth partner is to combine the Global Gateway with the European Green Deal, the EU's $1 trillion initiative for tackling climate change at home and abroad. Climate action in Africa is about greening the continent's heavy industries and developing new, climate-friendly business sectors that use the digital economy. These sectors of Africa's green economy could rely on the connectivity created by the Global Gateway: infrastructure for green energy production, as well as advanced information and communications networks. Therefore, the EU should use the Global Gateway to implement the international aspects of the European Green Deal. In fact, bringing Europe into Africa's green transformation would be beneficial for both continents (European Council on Foreign Relations 2021c).

The need of the hour is a *global power* willing to drive the worldwide effort to rebalance climate power and this is especially required at the COP26 Summit. Since the EU's own contributions to global emissions is comparatively small, it will need to play a vital facilitation role at COP26 and beyond if its own commitments are to make a difference to climate outcomes. The European Union has an opportunity to push other powers to implement the commitments they make at the all important COP26 Climate Change Conference. The EU certainly has the profile, experience and gravitas to assume this role and responsibility. As a powerful advanced economy and a major donor, as an important export market and regulatory superpower and over and above, as an essential source of expertise in intellectual property and infrastructure, the European Union can certainly assist other powers implement the green transition. With the announcement of the European Green Deal in 2019, the European Union has become the first global power to lay out concrete guidelines and proposals to reduce its greenhouse gas emissions by at least 55% compared to 1990 levels by 2030 (European Council on Foreign Relations 2021c). The EU is, therefore, uniquely positioned to implement the Green Deal and assume its responsibility as a significant global climate leader.

References

APN News (2021). 9 EU countries oppose market reform ahead of energy talks, accessed on October 28, 2021. https://apnews.com/article/business-europe-germany-spain-luxembourg-da0d2e67e9f07a3fb627144f17ae62cd.

Chatham House (2021). What is COP26 and why is it important? accessed on October 26, 2021. https://www.chathamhouse.org/2021/09/what-cop26-and-why-it-important?gclid=CjwKCAjwzt6LBhBeEiwAbPGOgbxkVYv5YW6EyCdQeBvWikZMQfKQNviwl_AXBSsROnDfYbddPRoCyTIQAvD_BwE.

Euronews (2021). Hungary's PM Orban blames EU climate change actions for energy price surge, accessed on October 27, 2021. https://www.euronews.com/next/2021/10/08/power-prices-eu-hungary.

Europa (2019). Delivering the European Green Deal, accessed on October 20, 2021. https://ec.europa.eu/info/strategy/priorities-2019-2024/european-green-deal/delivering-european-green-deal_en.

Europa (2020). The EU Green Deal Explained, accessed on October 20, 2021. https://www.nortonrosefulbright.com/en/knowledge/publications/c50c4cd9/the-eugreen-deal-explained.

Europa (2021a). Carbon border adjustment mechanism, questions and answers, accessed on October 26, 2021. https://ec.europa.eu/commission/presscorner/detail/en/qanda_21_3661.

Europa (2021b). EU at COP26 Climate Change Conference, European Commission, accessed on October 26, 2021. https://ec.europa.eu/info/strategy/priorities-2019-2024/european-green-deal/climate-action-and-green-deal/eu-cop26-climate-change-conference_en.

Europa (2021c). Strategy for financing the transition to a sustainable economy, accessed on October 26, 2021. https://ec.europa.eu/info/publications/210706-sustainable-finance-strategy_en.

European Commission (2019). European industrial strategy, accessed on October 22, 2021. https://ec.europa.eu/info/strategy/priorities-2019-2024/europe-fit-digital-age/european-industrial-strategy_en.

European Commission (2021a). Energy Prices: Commission present a toolbox of measures to tackle exceptional situation and its impacts. European Commission Press Release, Brussels, October 13, 2021, accessed on October 28, 2021. https://ec.europa.eu/commission/presscorner/detail/en/IP_21_5204.

European Commission (2021b). Farm to Fork strategy. accessed on October 24, 2021. https://ec.europa.eu/food/horizontal-topics/farm-fork-strategy_en.

European Commission (2021c). Renovation wave. accessed on October 22, 2021. https://ec.europa.eu/energy/topics/energy-efficiency/energy-efficient-buildings/renovation-wave_en.

European Council on Foreign Relations (2021a). 9 EU countries oppose market reform ahead of energy talks, 25 October 2021, AP News, accessed on October 28, 2021. https://apnews.com/article/business-europe-germany-spain-luxembourg-da0d2e67e9f07a3fb627144f17ae62cd.

European Council on Foreign Relations (2021b). The EU's Global Gateway and a new foundation for partnerships in Africa. *Policy Brief*, September 29, 2021, accessed on October 28, 2021. https://ecfr.eu/article/the-eus-global-gateway-and-a-new-foundation-for-partnerships-in-africa/.

European Union (2019a). Delivering the European Green Deal, accessed on October 20, 2021. https://ec.europa.eu/info/strategy/priorities-2019-2024/european-green-deal/delivering-european-green-deal_en.

European Union (2019b). A European Green Deal — Striving to be the first climate-neutral continent, accessed on October 20, 2021. https://ec.europa.eu/info/strategy/priorities-2019-2024/european-green-deal_en.

European Union (2020a). Proposal for a Regulation of the European Parliament and the Council establishing the framework for achieving climate neutrality and amending Regulation. 2018/1999 (European Climate Law), accessed on October 20, 2021. https://eur-lex.europa.eu/legal-content/EN/ALL/?uri= CELEX:52020PC0080,COM/2020/80final.

European Union (2020b). Climate action and the Green Deal. https://ec.europa. eu/info/strategy/priorities-2019-2024/european-green-deal/climate-action-and-green-deal_en, accessed on October 20, 2021.

European Union (2020c). Questions and answers: The revision of the Ten-E regulation, accessed on October 21, 2021. https://ec.europa.eu/commission/presscorner/detail/en/qanda_20_2393.

European Union (2021a). European Commission, Circular economy action plan. accessed on October 22, 2021. https://ec.europa.eu/environment/strategy/circular-economy-action-plan_en.

European Union (2021b). The EU Green Deal explained, accessed on October 21, 2021. https://www.nortonrosefulbright.com/en/knowledge/publications/c50c4cd9/the-eu-green-deal-explained.

Open Democracy (2021). A Green New Deal for whom? Alfons Pérez, accessed on October 27, 2021. https://www.opendemocracy.net/en/oureconomy/green-new-deal-whom/.

Reuters (2021). EU countries splinter ahead of crisis talks on energy price hike, accessed on October 27, 2021. https://www.reuters.com/business/energy/eu-countries-splinter-ahead-crisis-talks-energy-price-spike-2021-10-26/.

Stanly, J. (2021). Explained, Why natural gas prices are rising. *The Hindu*, October 9, 2021, accessed on October 27, 2021. https://www.thehindu.com/news/international/explained-why-natural-gas-prices-are-soaring/article 36910461.ece.

The Guardian (2020). What is the European Green Deal and will it cost €1tn? accessed on October 20, 2021. https://www.theguardian.com/world/2020/mar/09/what-is-the-european-green-deal-and-will-it-really-cost-1tn.

The Third Pole (2021). What is COP26, and why is it so important? accessed on October 20, 2021. https://www.thethirdpole.net/en/climate/cop26-explained/.

The Times of India (2021). EU Chief says key to energy crisis is pushing Green Deal, accessed on October 21, 2021. https://timesofindia.indiatimes.com/world/europe/eu-chief-says-key-to-energy-crisis-is-pushing-green-deal/articleshow/87156934.cms.

Weforum (2021). What you know about the European Green Deal — And what comes next, accessed on October 26, 2021. https://www.weforum.org/agenda/2021/07/what-you-need-to-know-about-the-european-green-deal-and-what-comes-next/.

Chapter 8

International Solar Alliance: Testing a New Framework to Approach Energy Shortage

Claudia Astarita* and Julius Hulshof*

Abstract

The International Solar Alliance (ISA) is a joint initiative of France and India that was launched during the Climate Conference in Paris in December 2015 (COP21) with the aim of making an unprecedented effort to promote solar energy to achieve three major goals: creating the conditions for a major decrease in the cost of solar energy; meeting the high energy demand in developing countries; and contributing to the fight against climate change. To achieve these objectives, ISA committed to set new ground rules, norms and standards for solar energy, regularly discussing with member countries the opportunities of developing innovative capacity-building measures and financial instruments to contribute to the harmonization of public policies, regulations, and prices between

*Dr. Claudia Astarita is lecturer at Sciences Po Paris. She obtained her Ph.D. in Asian Studies from Hong Kong University and her main research interests include China's political and economic development, Chinese and Indian Foreign policies, East Asian regionalism and regional economic integration. Julius Hulshof is an independent researcher with a dual masters in International Relations from Sciences Po and Peking University. His research interests range from China's socio-political development and foreign policy to East Asian international security, media and civil society.

the countries. ISA has been conceived as an action-oriented, member-driven, collaborative multilateral platform aimed at reducing users, governments, and investors' uncertainties related to the opportunity of relaunching both solar energy market and infrastructures to enhance energy security and sustainable development. Six years after ISA's foundation, this chapter aims at discussing both its major achievements and failures, as well as to assess the validity and the strength of its multilateral framework as a model for dealing with some of the major energy-related challenges in a post COVID-19 world. In particular, the chapter will discuss the ability of the ISA to add new dynamism and momentum to energy diplomacy in the 21st century, alongside the feasibility and advantages of creating a similar framework to facilitate a multilateral coordination in other areas, such as wind energy, and hydro energy.

Keywords: The International Solar Alliance, Solar Energy, India, Climate Change, Multilateralism, Indo-Pacific

Introduction

The deployment of renewable energy has made significant advances all across the world. For the last 3 years, renewable power investments have more than doubled those in fossil fuels (Ghosh and Chawla 2021). Especially in the developing world, investments in green energy are of great importance and should be considered a global concern because of the rising energy demand driven by quickly expanding developing economies. Driven significantly by India and China since the early 2000s, non-OECD countries altogether are estimated to account for over two-thirds of the global energy demand (Ghosh and Chawla 2021). In terms of people's consumption, there are over six billion energy consumers enjoying increasing living standards in today's developing world, nurturing estimations of a further 30% growth of energy demand over the next 15 years (Benoit 2019).

In order to meet the growing energy demand of a developing world currently transitioning to a low-carbon future, Indian Prime Minister Narendra Modi and French President François Hollande had, on the sidelines of the 2015 UN COP21 Climate Change conference in Paris, jointly launched the International Solar Alliance (ISA) (Osho 2020). The primary purpose of the ISA as an action-oriented, member-driven organization is to increase and scale up the use and quality of solar energy to meet the energy demands of solar-resource-rich countries in an affordable, equitable, and sustainable manner. More specifically, it is intended as a platform which brings together those countries with great solar potential, along with developers, solar innovators, and investors (India Global Business Staff 2018). The necessity of such a platform is easily understood considering that the 121 countries lying between the two tropics account for almost three-quarters of the world's population, yet only 23% of global solar capacity (Shidore and Busby 2019).

Despite being rich in solar energy resources, many of these countries remain still constrained by a lack of technology and capital. The proposition of ISA aims to address these challenges and promote the scaling up of solar energy applications by taking coordinated action aimed at better harmonization, aggregation of demand, risk and resources, for promoting *inter alia* solar finance, solar technologies, innovation, R&D and capacity building across countries (International Solar Alliance 2021). Besides ISA's flagship goal to mobilize investments of at least $1 trillion by 2030, the organization's central motive is to undertake collaborative and coordinated efforts to reduce the cost of solar finance while improving and making more affordable solar technologies through R&D activities.

To this day, developing economies receive too little renewable energy investments, and continue to face high costs of finance for solar-related projects (Ghosh and Chawla 2021). Solar markets tend to be too fragmented with governments lacking the technological insight and capability to differentiate and implement the policies and technologies necessary for investment in solar energy. While technology challenges must not be underestimated, these are arguably easier to solve than those of financial nature (Janardhanan and Chaturvedi 2021). Among the main aims of the ISA alliance there is the one of aggregating demand across countries to reduce the costs of solar technology as well as the cost of finance for the deployment of existing technologies (Ghosh and Chawla 2021).

The ISA makes clear its intent not to duplicate or copy the work of other energy organizations, such as the International Renewable Energy Agency (IRENA), the International Energy Agency (IEA), and the Renewable Energy and Energy Efficiency Partnership (REEEP). As stated by Ghosh and Chawla (2021) while referring to the many climate and energy technology partnerships emerging in recent decades, "there was seldom any actual transfer of technology, nor any explicit mandate to deploy advanced technologies". The global energy landscape appears to have been dominated by either groups of powers controlling the energy supplies (such as OPEC) or by energy-demanding OECD member states (such as the IEA), thereby excluding the new major energy demanders in developing countries and particularly, Asia (Ghosh and Chawla 2021).

Against this background, the ISA represents a truly multilateral effort to achieve the common ambition and goal of promoting solar energy and solar applications in all of its member countries, particularly those 121 "sunshine nations" lying between the tropics.

ISA: As an India-led Initiative

As the first treaty-based intergovernmental organization to be headquartered in India, Prime Minister Narendra Modi has announced the alliance as India's "gift to the world" in the battle against climate change (Amaresh 2021). Due to its leadership role alongside its generous contributions to the organization, India is without doubt the most indispensable member country to the existence of ISA and intends, grounding on domestic energy efforts, to serve as an example to other developing countries. It has allocated land for the construction of the organization's headquarters together with $27 million for infrastructure construction, while creating a corpus fund and supporting all costs of ISA Secretariat for the first five years (Ghosh and Chawla 2021).

As one of the fastest growing economies on Earth and a population expected to rise to 1.6 billion by 2040, and thereby possibly quadrupling the nation's energy demand, it is of vital importance for India to alter its energy mix and invest in renewables and natural gas, coal and oil to provide electricity and keep up with the rising living standards of its citizens (Sivaram 2017). This context explains India's leadership role in the sector of solar energy as well as its domestic track record over the past decade. Next only to China and the United States, India has the world's third fastest expanding solar power program and an ambitious energetic agenda. After meeting the 2022 aim of 20 gigawatts of solar capacity four years before the set deadline, Prime Minister Narendra Modi has announced India's new target as the one of increasing its solar capacity to 100 gigawatts by end of 2022 (*Reuters* 2015).

The solar agenda of Modi's government includes the "24x7 Power for All", i.e., to provide electricity to all Indian households, installation of numerous grid-scale solar parks and subsidies to install solar-powered water pumping systems for irrigation and drinking water with which to empower rural populations (Purushothaman 2018). Other notable achievements in the field of solar energy are the Indian railways trains with solar rooftops to provide electricity inside the coaches, and the seventh busiest airport in India, Cochin International Airport located in the city of Kochi, being run entirely on solar power. Further, the costs of solar energy have been progressively decreasing, thanks to renewable energy bids sold at lower rates than the coal power generation (Silvio 2018; Buckley 2017).

However, in spite of India's impressive solar scale-up in ISA, it still faces considerable obstacles to realize its renewable energy capacity goals and to make the use of solar energy possible across all of its regions. Domestically, some states face a variety of challenges — ranging from the lack of available land, opposition of established interest groups, and limited political will — while the development and investment landscape of solar energy is equally imperfect with, among other issues, the monopsony power of federal state-owned electricity distribution companies (Shidore and Busby 2019). Regardless of such challenges, rise in India's solar consumption has provided the developing world with many useful lessons, granting it the role as a solar leader. While enhancing its status as a globally significant player in the push to tackle climate change challenge, India and its Prime Minister Narendra Modi appear to be legitimate figurehead for the promotion of solar energy cooperation and investment in the developing world and beyond.

ISA Membership and Institutional Characteristics

The ISA is a member-driven organization initially designed for the 121 "sunshine countries" lying fully or partially between the two tropics only. In December 2017, the ISA formally became a treaty-based international intergovernmental organization when 15 states had ratified its Framework Agreement. In 2018, the alliance opened its membership to all United Nations member states to put solar energy on the global agenda, and create a more inclusive and universal appeal for the development of solar energy (Kabeer 2018). Currently, 80 countries, from all across the world, have ratified the Framework Agreement of the ISA, though the organization is said to have at least 121 prospective member countries — those with national territories inside the two tropics — while aspiring to achieve universal membership (International Solar Alliance 2021).

Jointly launched by India and France in 2015, the United Kingdom was the second European country to ratify the ISA Framework Agreement in 2018, with the British International Development Secretary Penny Mordaunt stating that "the signing of this treaty is a momentous occasion for the UK, demonstrating our continued commitment to providing the very best of British expertise to the renewable energy sector" (Department for International Development and The Rt Hon Penny Mordaunt MP 2018). After the Netherlands became a member country in 2018, Germany, Sweden, and Denmark ratified the Framework Agreement in 2021 (*The Economic Times* 2021). As stated by its government ministries, "Germany will be able to play an active role in ensuring that the young organization develops its full potential in the best possible way in cooperation with other international energy organizations". While China and the United States — both countries with great technology and production capacity might — are not yet members of the ISA, both are considered prospective member countries.

The potential membership of the United States and People's Republic of China to the ISA has been evidently affected by the ever-present geopolitical elements shaping interstate equations and interactions on international platforms. An alliance spearheaded by India may not be warmly welcomed by China and Pakistan for fear of losing regional influence, while the United States could perceive any form of emerging multilateralism in the Asia-Pacific region as undesirable. Countering such notions, the ISA appears to receive the support and trust from its member countries as it employs a framework that does not intend to impose anything beyond cooperation in the field of solar energy through a truly multilateral agenda.

Since its formal inception as an international organization, the ISA operates by means of a yearly Assembly, the decision-making body comprising the representatives of each member state functioning by means of several committees, alongside a Secretariat purposed to assist all their actions, projects, and activities by members. When it comes to the institutional features of the alliance, the fundamental principles laid out in the Framework Agreement demonstrate its very multilateral nature. For example, any program must be proposed by at least two member countries or a larger group of members, while a proposal can be adopted if supported by two members without objections by more than two member states (*International Solar Alliance* 2021). In this way, only programs in which several countries find common interest will be launched, thereby avoiding the risk of certain countries to singlehandedly dominate the alliance's agenda.

Further, the ISA adopts a model with relatively little member commitments, meaning that it does not consist of binding pledges, membership fees, or mandatory state funding (Shidore and Busby 2019). This appears to have positively contributed to the membership of countries such as the

Netherlands, as the Dutch minister of foreign affairs, Sigrid Kaag, defended the 2018 Dutch accession to the alliance while referring to the absence of any costs accompanying membership; especially with host country India offering to cover the organization's costs until 2021 (van Gastel 2019). Listed in Article VI of the UN Framework Agreement, ISA's budget is to be covered by voluntary contributions by its members, partner countries, and the UN, as well as from the private sector (International Solar Alliance 2021).

The role of the Secretariat, beyond its ability to propose new programs and initiatives, is to ensure coherence among the ISA agenda and programs altogether; all decisions about the implementations of programs are taken by the member countries which have autonomously decided to actively contribute to their implementation. The action-oriented nature of the ISA's organizational model is reflected in the relatively small secretariat which is supposed to bring together the right set of actors in accordance with each and every work programs. As stated by Ghosh and Chawla (2021), it is essential for the secretariat to be strong in diplomacy and clear in its communications while inspiring confidence "by the mastery of content related to solar technologies, policies, finance and markets".

Some examples of the alliance's multilateral character, and the move away from it being an India-centered organization can be found in its 2025 deliverables from the ISA 2020 *Annual Report* (International Solar Alliance 2021). Here, the universal nature of its membership, for all the UN members and the recruitment of global staff members to create a multilateral Secretariat are perhaps most notable. The intention for the alliance's financial status to no longer solely depend on India is also clear, as it intends to embrace financial sustainability to new avenues including membership fees and host country grants.

ISA's Programs and Progress

In its six years of existence, ISA's expansion, efforts, and global reach thus far have been impressive, as its vision statement in the 2020 Annual Report outlines its mission saying, "In the true spirit of multilateralism, the ISA as the fastest growing body enabling energy transition; is the solution the world has long enough waited for" (ISA Annual Report 2020). In line with being the fastest growing body enabling the energy transition, the organization has not only grown its institutional presence but also already launched a number of work programs since 2016. These include, at present: scaling solar applications for agricultural use; affordable finance at scale; scaling mini-grids; scaling solar rooftop, and scaling solar e-mobility and storage (International Solar Alliance 2021). While this section does not intend to provide an exhaustive account or chronology of the ISA's journey thus far, it purposes to give a general idea of its activities and progress to date.

To start with, the program on "Scaling Solar Applications for Agricultural Use" (SSAAU) focuses on the deployment of solar water pumping systems as a solution to farmers in member countries for energy access and sustainable irrigation. To make such projects feasible, the ISA secretariat has aggregated various countries' demand to reduce system costs and accordingly undertaken a global tendering process for international competitive bidding for all project-related services (International Solar Alliance 2021). The ISA has also included various country missions to better grasp the challenges and issues on ground-level, gaining insights critical to the success of the solar water pumping program. While these missions and this working program have invited the interest and collaboration of many local and international stakeholders, the lack of affordable finance appears to have remained the major obstacle to the actual adoption and deployment of solar water pumping system technology.

When it comes to the "Affordable Finance at Scale" program, the Solar Risk Mitigation Initiative (SRMI) has been launched at the COP24 by the World Bank and the Agence Francaise de Développement (AFD) to support the key aims of the ISA: to leverage private sector investment for the development of bankable solar programs in developing countries (ISA Annual Report 2019). The World Bank has, in partnership with ISA, committed $337 million for those African member countries in the off-grid sector, while the European Investment Bank (EID) has initiated a €60 million grant investment project to promote off-grid applications in Africa. Other institutions such as the Export Import Bank of India, France's AFD and the Asian Development Bank have approved great sums of money to support respective programs for ISA member countries.

These funding patterns supporting the deployment of solar energy go hand in hand with ISA's financing roadmap for mobilizing $1 trillion by 2030. This impressive roadmap and consequent action plans specific to the ISA member countries have been developed in a joint mission with representatives from various multilateral development banks and development finance institutions, as well as leading global consulting and research firms (all partner organizations to the ISA). As a knowledge aggregator, the ISA, among other projects, also works on the "Scaling Solar Minigrids", the "Scaling Solar Rooftop", the "Scaling Solar E-mobility & Storage", the "ISA Solar Park", and "Solarizing Heating and Cooling Systems" programs; all of which form early stages to move onward by receiving significant support from member countries, resulting in various country missions, implementation plans, and technical reports. The ISA has aggregated the total energy demands per project, with as many as 53 countries submitting demand for the "Solar Home System" project. While most of these programs are still under construction, the progress is palpable with, for example, the ISA bringing down the price of nearly 300,000 solar water pumps by almost half through a global price tender. Additionally, it has concluded 10 of its expert country missions out of the 15 targeted countries.

The alliance is additionally proactive through other endeavors of which the ISA Solar Technology and Application Resource Centre (ISA STAR C) presents an apt example, alongside the successfully launched "Infopedia" — an online platform to disseminate information and knowledge on solar energy. It also provides the extensive Indian Technical and Economic Cooperation (ITEC) driven training program to various participants from ISA member states, while offering the ISA Solar Fellowship for mid-career professionals. This scheme currently has 21 candidates from 18 member countries pursuing studies to eventually contribute to solar energy management and economics in their home countries. Finally, four different ISA awards have also been institutionalized to further promote the awareness of the alliance and the developments in the solar energy sector.

ISA's Multifaceted Partnerships

Alongside and embedded in the programs and projects of the ISA are its many partnerships with major global financial, political, and environmental institutions. At the 2018 Founding Conference of the ISA, for example, the ISA had signed a deal with the African Development Bank (AfDB) to scale up solar energy in Africa through technical assistance, knowledge transfer and the development of innovative financial instruments to reduce risks and lower the costs associated with solar investments (African Development Bank, 2018). As part of this initiative, the ISA supports — among other of the AfDB's initiatives and its leadership role in the "New Deal on Energy for Africa" — the AfDB's "Desert to Power solar initiative" which will deploy 10,000 MW of solar power systems in Africa's deserts in order to provide electricity to 250 million people, a great part of which will be off-grid access.

Then, the ISA has also signed a joint financial partnership declaration with the European Bank for Reconstruction and Development (EBRD) to deepen cooperation in support of renewable energy (International Solar Alliance 2017). Declarations such as these and the ISA's earlier and first financial partnership with the World Bank have paved the way for a great deal of evolving new collaborations with various multilateral and development banks, all in order for it to achieve its mandate and to ease the cost and affordability of solar energy. Institutions such as the EBRD tend to be keen to collaborate with the ISA in order for them to increase their green finance portfolio and contribute to the global vision of leading the global economy through sustainable development and green energy. As a result, various major development banks have guaranteed to pledge funds, with the Asian Development Bank, for example, committing $3 billion per year by 2020 (Fitch 2018).

Aware that each program and partnership deserves more elaborate analysis and description, and despite not having delved into each and every ISA action, the above representative overview perhaps most amply demonstrates the broad range of ISA's activities and progress characterizing the nature and functioning of this organization. Especially, its multidimensional focus on energy development becomes even more apparent when reading through the organization's annual reports or when viewing the 30 plus key partner organizations with which the ISA has initiated various collaborative partnerships in solar energy applications. Most importantly, perhaps, is the significant interest and participation of member states (particularly from the so-called "sunshine nations") that has seen robust implementation of various ongoing work programs and activities. So far, the ISA's endeavors appear to comply with its description as a specialized and action-oriented intergovernmental organization.

The Alliance's Future Initiatives

In an interview with Indian daily newspaper *LiveMint*, the ISA's Director General, Ajay Mathur, had recently delved into the future vision of his organization wherein the World Solar Bank and the One Sun One World One Grid (OSOWOG) initiative features most prominently among others. These two initiatives aim to integrate the long-term goals of the Alliance together with the many other projects and programs that have been implemented or initiated by various member states. Both initiatives were announced at the COP26 in November 2021 (Bhaskar 2021).

The "One Sun, One World, One Grid" initiative, articulated by Indian Prime Minister Narendra Modi, essentially revolves around the idea that "the sun never sets", as it is constantly shining at a geographical location globally at any given time. With ISA member countries located across all continents, OSOWOG's fundamental concept is the development of "a trans-national grid that will be laid all over the globe to transport the solar power generated (and harnessed) across the globe to different load centres" (*International Solar Alliance* 2021). In this global grid, the large-scale solar generation capacity from certain places and regions would, through bi-lateral, regional, and interregional transmission connections eventually lead to global interconnection of solar energy resources with solar energy transferred where it is needed.

In the first ever World Solar Technology Summit on September 2020, organized by the ISA and Government of India, a Memorandum of Understanding (MoU) regarding the OSOWOG initiative was signed between the ISA, the Government of India, and the World Bank. According to this MoU, the alliance is expected to act as the nodal agency for all activities surrounding the OSOWOG. Among others, this includes undertaking a profound study covering the entirety of the program's implementation. Funded by a sustainable partnership between the World Bank and the State Bank of India, the OSOWOG initiative is conceived on three different phases of implementation, starting with the Indian grid's interconnection with neighboring Asian regions, then to the African continent, and finally the rest of the world. Also, the initiative has been presented and promoted as instrumental to achieve the mitigation targets set out in the Paris Agreement (International Solar Alliance 2021).

With the ball set to roll, the road map of OSOWOG's implementation is anticipated to be ready by end of 2022, and the requisite technology for storing and transmitting electricity established and proven to work. What remains to be confirmed thereafter is the global political will to strive for such sustainable interdependency and the consequent mobilization of financial means to accomplish this inordinate goal. To address the latter, alongside the OSOWOG as a key ISA deliverable at the COP26 in Glasgow was the initiative of the World Solar Bank, a new institutional partner that aims to address solar energy-related financial issues (Bhaskar 2021). Meanwhile, the aggregation of demand for solar projects mentioned, requiring major investments in solar pumps, solar rooftop, solar mini-grids, solar parks, and solar home systems, remains premised on the availability of financial requirements that goes well beyond the capabilities of many member countries put together. The main difficulty of mobilizing finance of such magnitude typically revolves around the risks associated with renewable energy investments in developing countries due to factors such as the lack of policy framework and regulations (Ghosh and Chawla 2021).

This is where the ISA plans to set in to establish a World Solar Bank as a financial institution to fund such mega solar projects among these member states. Addressing these challenges of facilitating flow of funding into such large-scale solar investments is not an easy task, and as Director General Ajay Mathur noted: "This is again a work in progress because we would like to make sure that all issues — whether it is risk management, policy management, coordination of financial investments, development of DPRs (detailed project reports) that are bankable — get addressed" (Bhaskar 2021). With the challenges of the high cost of finance and technology inherent to nearly all efforts to scale up the deployment of solar technology in developing countries, the proposal of a World Solar Bank to achieve the flagship goal of mobilizing US$1 trillion by 2030 appears an uphill and yet a sensible task. One the eleven 2025 deliverables noted in the ISA annual report of 2020 was to support the World Solar Bank in mobilizing 50 billion a year through co-financing of projects (ISA Annual Report 2020). Though it is yet unknown where the bank will be headquartered, it has been noted that the host country needs to finance 30% of the proposed capital size of $10 billion. The proposal for establishing this novel global financial institution was expected to be taken to the next ISA Assembly and its Standing Committee for guidance and advise to be launched at the Conference of Parties (COP26) in Glasgow (Tandon 2021).

Challenges before ISA

The success of the ISA currently depends on member states' capacities to overcome numerous potential challenges; economic, political, and socio-cultural. With New Delhi at the forefront of this Alliance's goals as a solar leader, it is essential for India to have a good track record and achieve progress in its domestic solar program and projects. Shidore and Busby (2019) explain that as a key founder of this Alliance, New Delhi's ability to lead domestically and internationally while helping the ISA achieve its agenda will be crucial for the organization's health and international status and success.

Second, some analysts argue that for the ISA to function as effectively and meaningfully as possible, more developed countries must join the Alliance. This would especially require the largest consumers of energy as also world's largest polluters, namely the United States and China, be co-opted into the Alliance. In addition to their own drive to reduce their enormous carbon footprint, their technological advantage, together with their financial and manufacturing capacity plus their current dominant position in the solar supply chain, will be decisive for ISA success from two different perspectives. A purely energetic one would mean an increment to ISA capacity to produce and distribute solar energy, and second, from a geopolitical perspective, it would increase ISA's options for multilateral dialogue encouraging most other states to join ISA's work on pragmatic and mutually beneficial goals of clean energy consumption.

Third, the ISA novel governance structure especially deserves some serious consideration. The Alliance has been founded on a model that, on the one hand, limits member countries' commitments and, on the other hand, emphasizes on the importance of private sector investments especially from the corporate sector. It is argued that with States' tendency to over-bureaucratize and micromanage that sometime is seen characterizing India's projects, ISA future institutional development must be carefully followed in letter and spirit of ISA's ambitions (Shidore and Busby 2019). For this reason, other countries should be encouraged to play an active role in orienting ISA toward a more adaptable, flexible, and lean structure. The earlier-described initiatives for the ISA that seek to internationalize its initiatives by increasing its global membership aim to avoid attaching it too firmly to the centrality of India and New Delhi's approach so far appears encouraging in this regard.

Fourth, even though the crux of the Alliance is to work with and bring together various actors from both the public and private sectors in international arena, it must ensure these partnerships are built as equal partnerships to make them effective. While recognizing the necessity of outsourcing ambitious financial aims to private actors and institutions like the World Bank, it is important to acknowledge that these actors inevitably have an agenda of their own as well. Communication among ISA member states and other stakeholders, therefore, becomes an essential part of its function, especially so in articulating its priorities in the partner-driven ecosystem with multilateral banks, financial institutions, involving a whole range of other actors, think tanks, UN bodies, and private and public institutions. To this end, the more member states the ISA has the more active will their participation in the organization, the more collective the ISA will be and thereby it will be able to provide for the needs of its members and develop stronger leverages as also jurisdiction to exert internationally.

At the same time, since this novel model of the Alliance seeks to promote vibrant and productive collaborations with actors from across the globe, primarily due to its specialized and action-oriented nature, this member-driven foundation, the more "solar energy oriented" focused it remains, the more pragmatic and realistic its efforts will be, and more easily will it be able to generate binding partnerships while attracting number of new member states to its fold. ISA's impact and effectiveness would benefit a lot from remaining focused on three priorities highlighted by the French President Emmanuel Macron in 2018: to identify solar projects, to mobilize public and private finance at scale with a focus on guarantee instruments, and to transfer innovative technology solutions and capacity building (African Development Bank 2018).

As Ajay Mathur, the Director General of the ISA, argues, ISA's agenda is essentially political and yet it remains driven by the common global goal of economic growth with climate change mitigation which makes moving toward solar so important (Bhaskar 2021). This common goal is expected to bring all countries together in this global Alliance. As he says, "it is about all countries coming together to help provide whatever it takes — whether it is experience, expertise, resources or the ability to adopt solar technology" (Bhaskar 2021). The message of ISA's Director General is clear, as he pronounces the ISA to be "delighted" if all UN countries including China, Pakistan, and the United States were to join the Alliance.

Transformative Power of the ISA

To deal with the world's growing energy demand while addressing the large-scale environmental problems and the need to use clean energy, the move toward renewable energy as an alternative energy source becomes an even more apparent prerequisite for human survival. In line with the ISA's ambitions, solar energy is expected to play an important part in carrying these two simultaneous transitions in the global energy sphere and its potential in the domain of energy security cannot be overstated. Particularly for those emerging economies located in between the Tropic of Cancer and the Tropic of Capricorn, tapping the uninterrupted access to sun rays for electricity generation will guarantee a regular improvement of life standards offering a chance to meet local energy needs at local level (Mohapatra 2019). Solar energy has the potential to relieve developing countries from the difficulties accompanying conventional energy sources such as oil, gas, and coal, because of the ability to harness and produce solar energy locally, at minimal costs; meaning that countries could be relieved from the stress and disquiet to afford expensive oil- and gas-based energy from those specific locations around the world where this is available.

It also has critical implications for global geopolitics and international relations. Free of charge and abundantly available, the significance and geopolitical consequences of a wider deployment of solar energy could be momentous as energy-dependent countries will be able to achieve energy security by themselves. It is a matter of fact that a great deal of the geopolitical conflicts in the 20th century were caused by energy needs, while growing radicalization and forms of religious extremism in different parts of the world are also oftentimes tied to energy geopolitics and the support of these groups by oil-producing states (Mohapatra 2019). Exponentially rising demand for energy indeed is expected to be a major trigger for future conflicts around the world. If solar energy will be deployed more efficiently and widely on a global scale then dependence on oil-producing states might decrease and, expectedly, the likelihood of oil interest-fueled conflicts can be contained.

According to some analysts, the ISA could also nurture a more just international energy order by giving equal representation to smaller and poorer states such as Fiji and South Sudan alongside those advanced economies such as Australia and Germany (Ghatak 2021). During the first ISA assembly in 2018, Prime Minister Narendra Modi expressed his optimism about the organization becoming an "alternative OPEC", a view given greater weight when Saudi Arabia also became a member country to the Alliance in 2019 (Mohapatra 2019). Generally, it is the ISA's alternative governance that promises to enable it to navigate through and potentially reconstruct regional balances of power through mutually beneficial partnerships irrespective of any geopolitical contestations. Numerous forms of collaborations have already started to emerge, including ISA and numerous regional groupings such as the Shanghai Cooperation Organization, the Group of Latin America and Caribbean Countries, and the AfDB. Moreover, transregional bilateral cooperation, such as that between India and Bolivia, where both countries agreed to jointly develop Lithium batteries — of which the largest deposits exist in Latin American countries — after Bolivia expressed interest in joining the ISA, also indicates the win–win potential of the solar energy collaboration (Mohapatra 2019). This novel type of cooperation through an organization spearheaded by the developing world appears to grant the alliance its sway as it more easily gains the trust and support of developing countries facing similar energy problems. Finally, the ISA's multilateral and cooperative nature are also reflected in programs such as the "One World, One Sun and One Grid" initiative and the World Solar Bank.

To conclude, though it may be yet hard to assert how and to what extent the ISA and its member countries will be able to further strengthen their collaboration and reduce their energy dependencies on oil-rich countries, this alliance certainly shows the potential to add new dynamism and momentum to energy diplomacy in the 21st century. Steering toward universal membership, the ISA has the potential to become a major platform geared toward the usage of renewable energy sources on an equitable basis by creating an "alternative energy grid" benefiting each and every state involved. Beyond geopolitical tensions, power politics and global instability, is where the ISA presents itself as a new age multilateral platform geared toward the single focal point that is solar energy. The promise of an aggregated solar energy market has already attracted attention and interests of many countries since the ISA's launch in Paris in 2015, and there is ample reason for the Alliance to draw more countries in this global movement toward sustainable growth and to play as a referent for similar alliances in other areas.

References

African Development Bank Group (2019, February 5). African Development Bank signs deal with International Solar Alliance to propel solar development in Africa. African Development Bank — Building today, a better Africa tomorrow. https://www.afdb.org/en/news-and-events/african-development-bank-signs-deal-with-international-solar-alliance-to-propel-solar-development-in-africa-17936.

Amaresh, P. (2021). India's "International Solar Alliance" in the post-COVID world. *Diplomatist*, January 29. https://diplomatist.com/2021/01/29/indias-international-solar-alliance-in-the-post-covid-world/.

Benoit, P. (2019). Energy and development in a changing world: A framework for the 21st century. *Columbia: SIPA Center on Global Energy Policy*. https://www.energypolicy.columbia.edu/research/energy-and-development-changing-world-framework-21st-century.

Bhaskar, U. (2021). ISA will be delighted to have China, Pakistan as its members: Ajay Mathur. *Mint*, March 21. https://www.livemint.com/news/india/isas-goals-are-political-director-general-ajay-mathur-11616317163868.html.

Buckley, T. (2017). IEEFA Asia: India's electricity-sector transformation is happening now. *Institute for Energy Economics & Financial Analysis*, May 17. https://ieefa.org/ieefa-asia-indias-electricity-sector-transformation-happening-now/.

Department for International Development and The Rt Hon Penny Mordaunt MP (2018). UK joins International Solar Alliance to help provide over 1 billion of the world's poorest people with clean, affordable energy. GOV.UK, April 16. https://www.gov.uk/government/news/uk-joins-international-solar-alliance-to-help-provide-over-1-billion-of-the-worlds-poorest-people-with-clean-affordable-energy.

Fitch, C. (2018, April 28). International solar alliance: Turning on the lights. *Geographical Magazine*. https://geographical.co.uk/nature/energy/item/2697-turning-on-the-lights.

Ghatak, D. (2021). Taking stock of the International Solar Alliance and its geopolitical implications. *Asia in Global Affairs*, July 26. https://www.asiainglobalaffairs.in/reflections/taking-stock-of-the-international-solar-alliance-and-its-geopolitical-implications/.

Ghosh, A. and Chawla, K. (2021). The role of international solar alliance in advancing the energy transition in Asia. In Janardhanan, N. and Chaturvedi, V. eds., *Renewable Energy Transition in Asia*. Singapore: Springer, pp. 63–87.

India Global Business Staff. (2018). The International Solar Alliance: From promise to action. *India Global Business*, November 16. https://www.indiaglobalbusiness.com/igb-archive/the-international-solar-alliance-from-promise-to-action-india-global-business.

International Solar Alliance Secretariat. (2019). ISA Annual Report 2019. https://isolaralliance.org/uploads/docs/f661c80f9c697a7645164b48fb9003.pdf.

International Solar Alliance Secretariat. (2020). ISA Annual Report 2020. https://isolaralliance.org/uploads/docs/22ea89a88f2b407da6c17d44f94cb7.pdf.

International Solar Alliance (ISA) and the European Bank for Reconstruction and Development (EBRD) sign Joint Financial Partnership Declaration to deepen the cooperation in support of Renewable Energy (2017). Press Information Bureau Government of India, Ministry of Finance. https://pib.gov.in/newsite/PrintRelease.aspx?relid=173169.

International Solar Alliance. US, China keen to join International Solar Alliance. *The Economic Times*. (n.d.), accessed on September 18, 2021. https://economictimes.indiatimes.com/news/politics-and-nation/us-china-keen-to-join-international-solar-alliance/articleshow/63259131.cms?from=mdr.

Janardhanan, N. and Chaturvedi, V. (2021). Framing the renewable energy context for Asia. In Janardhanan, N. and Chaturvedi, V. eds., *Renewable Energy Transition in Asia: Policies, Markets and Emerging Issues*. New York: Springer, pp. 3–17. https://doi.org/10.1007/978-981-15-8905-8_1.

Kabeer, N. (2018, November 2). ISA opens its membership to all United Nations member countries. *Mercom India*. https://mercomindia.com/isa-opens-membership-un-member-countries/

Mohapatra, N. K. (2019). Why the International Solar Alliance is geopolitically significant. *Down to Earth*, April 19. https://www.downtoearth.org.in/blog/energy/why-the-international-solar-alliance-is-geopolitically-significant-64080.

Osho, Z. (2020, May 22). Can the International Solar Alliance truly be India's "gift to the world"? *The Diplomat*. https://thediplomat.com/2020/05/can-the-international-solar-alliance-truly-be-indias-gift-to-the-world/.

Purushothaman, C. (2018). India's rising stature as a solar power. *The Diplomat*, March 14. https://thediplomat.com/2018/03/indias-rising-stature-as-a-solar-power/.

Reuters (2015). India's Modi raises solar investment target to $100 bln by 2022, January 2. https://www.reuters.com/article/india-solar-idUSL3N0UG13H20150102.

Shidore, S. and Busby, J. W. (2019). One more try: The International Solar Alliance and India's search for geopolitical influence. *Energy Strategy Reviews*, 26, 100385. https://doi.org/10.1016/j.esr.2019.100385.

Silvio, M. (2018, January 30). India coal power is about to crash: 65% of existing coal costs more than new wind and solar. *Forbes*. https://www.forbes.com/

sites/energyinnovation/2018/01/30/india-coal-power-is-about-to-crash-65-of-existing-coal-costs-more-than-new-wind-and-solar/.

Sivaram, V. (2017). Can India save the warming planet? *Scientific American*, May 1. https://doi.org/10.1038/scientificamerican0517-48.

Tandon, T. (2021, February 25). *What is World Solar Bank to be launched by International Solar Alliance (ISA)?* Jagranjosh.Com. https://www. jagranjosh.com/general-knowledge/what-is-world-solar-bank-to-be-launched-by-international-solar-alliance-isa-1614258823-1.

The Economic Times (2021). Sweden happy to join the International Solar Alliance: Envoy, July 23. https://economictimes.indiatimes.com/industry/ renewables/sweden-happy-to-join-the-international-solar-alliance-envoy/ articleshow/84675363.cms?from=mdr.

van Gastel, E. (2019). Solar Magazine — Kamervragen over International Solar Alliance: Deelname kost Nederland geen geld, April 11. https://solar magazine.nl/nieuws-zonne-energie/i18221/kamervragen-over-internationale-solar-alliance-deelname-kost-nederland-geen-geld.

Chapter 9

Multilateralism Efforts in Asia: What's the Way Forward?

Prathit Singh*, Namit Mahajan*, and Ritvick Khanna*

Abstract

Over the past decades, the world has seen persistent attempts by inter-national players to scrutinize a global consensus for acting on climate change, one of the most impending issues of the decade. Beginning at Copenhagen, being ill-sustained at Kyoto, and finding renewal at Paris, the climate change fight has shown us close moments wherein the inter-national community expected a final climate agreement, with defined limitations on carbon emissions, investments, and pollution control. However, these moments failed to yield any tangible outcome; with there being further divisions among international stakeholders. The incentive to reach a global consensus-based regime toward mitigation of climate change and its effects, thus far, has found inadequate actualization due to conflicts of interest between developing and developed nations. The need for a consensus at such a large international stage has contributed to many of its failures owing to the obsession with multilateralism. Such an approach has only led to a cycle of disruptions and no change what-soever. This chapter attempts to explore the concept of multilateralism; a significant contributing factor to futile agreements. Keeping the devel-oping world at the center of the study, this chapter primarily explores

*Students of Ramjas College, University of Delhi, India.

three questions. First, we examine the ideological and theoretical flaw in the multilateral obsession that leads to a continuous hindrance of the cause of the developing and the developed world. Then, we analyze the conflict of knowledge and expertise, which the Global South faces while confronting the Global North at the negotiation table. Finally, we investigate the structural flaws in the deliberation process guided by a multilateral idea of negotiations. Subsequently, this chapter also examines the alternatives to the multilateral idea while exploring minilateral methods with a deliberative democratic approach.

Keywords: Climate Change, Multilateralism, Minilateralism, Climate Negotiations

Introduction

The world has relied on multilateralism and multilateral institutions to solve what is perhaps the gravest problem of this century — climate change. However, years have passed since international communities' determination and resolution toward having a multilateral binding regime to limit the impacts of climate change, but to no avail. Pledges have been made, multilateral forums have been created, alliances have been formed with a common goal, but nothing seems to have changed apart from the rise in temperatures and increase in greenhouse emissions. Perhaps this inefficiency is somewhere attributive to the structural institution of multilateralism itself. What follows in this chapter is an analysis of multilateralism and its tenets which have formed and guided the institutions of climate change regimes. We intend to portray a detailed and in-depth analysis of the areas which have led to the failure of multilateralism as an institution. Philosophical, theoretical, and critical assumptions have been bases of our analysis. An attempt has been made to understand what the failures of multilateral institutions are and what factors have added to the failure to reach a negotiated deal. Some crucial aspects of multilateralism in the environmental regime have been challenged through this chapter.

The Tenets of Multilateralism

Before we delve into an analysis of the effect that multilateralism has had on the negotiations pertaining to climate change, it is important to define what our view of multilateralism is. Multilateralism as John Ruggie defines it in its ideal type is "an institutional form which coordinates relations among three or more states on the basis of 'generalized' principles of conduct" (Ruggie 1992:561). As an institution, multilateralism rests on recognized principles of sovereign equality, indivisibility, and diffuse reciprocity. The institution form of multilateralism in its ideal type invokes a Westphalian idea of a nation-state engaging with other nation-states with the principle of sovereign equality at helm, which implies that all conditions of a particular arrangement must apply equally to all parties to it, and since sovereignty serves as a high principle on the basis of which the arrangement is drawn, states are neither forced to comply nor are hesitant to withdraw. Not surprisingly, Ruggie's ideal type of multilateralism is almost never followed in the international domain, even by the powerful states which often lead the formation of these multilateral institutions. More so, with the post-colonial emergence, multilateralism as a principle has undergone serious evolutions. As Eckersley puts it:

"...it might be more fruitful to think of multilateralism not as a modern, generic institution with one set of constitutive principles, against which there are many exceptions, but rather as a much more flexible institution, with different principles and decision-making practices in different domains of governance" (Eckersley 2012:24).

As a principle, the evolution of multilateralism began with a hierarchical order wherein the powerful states dictated the terms of multilateral institutions to the less powerful states. Gradually, the idea evolved to mean formal equality in an institutional domain between the more powerful and the less powerful states. The latest form of the evolution of multilateralism, as has been seen in the environmental domain, embraces the idea of reverse discrimination to achieve substantive equality (Eckersley 2012:24). In the domain of negotiation of a treaty for climate change, a crucial example of this evolution, as termed by the UNFCCC, is the idea of "common but differentiated responsibilities" which implies that multilateral environmental regimes are developed on the basis of asymmetric and varying objectives on the developed states as a form of positive discrimination against the inequalities met by the developing states. Unlike ideal multilateralism, these regimes are not based on the idea of common objectives, equal sovereignty or reciprocal outcomes (Eckersley 2012:24). It is imperative to examine how this principle evolved in the formation of environment regimes.

Environmental multilateralism, which found its roots at the Stockholm Conference on the Human Environment, in 1972 recognized that the developing countries bear development needs and called for financial and technological assistance for addressing these needs (Eckersley 2012:24). By the time Montreal Protocol 1987 was signed, the idea of differentiated responsibilities had taken proper shape. It involved the "grand bargain" of ensuring Southern participation in the agreement in return of Northern Assistance (Eckersley 2012:24). This very principle is enshrined in Article 3(1) of the UNFCCC, which imposes an obligation on developed countries to "take the lead in combating climate change and the adverse effects thereof" (Article 3(1)). "This obligation is derived from their significant historical and current emissions, their high per capita emission and their greater capabilities (technological, economic and administrative) to pursue mitigation and assist developing countries with mitigation and adaptation" (UNFCCC 1992). This step is a form of affirmative action to reduce the asymmetries of power, capacity, and wealth between the developing and the developed world.

Affirmative Multilateralism and the United States of America Inc.

Even though this idea of affirmative action incorporated in the multilateralism of the climate regime proves to be in line with justice and informal equality, this principle has not been to the appeasement of arguably the most important actor in the regime; the United States of America. One of the causes that have led to the repeated failures of most of the climate agreements beginning from Kyoto has been the inconsistent participation of the United States. While on one hand, blocs like the European Union have been welcoming of the affirmative nature of the climate regime, the United States seeks problems in leaving the developing countries out of any binding obligation. The root of the problem is also ideological.

As Kagan points out, the root of ideas of blocs like the European Union lies in the Kantian ideal of eternal peace and focuses on developing a rules-based world. Instances and reflection of these ideas can be spotted in the development of the European Union Framework, the common currency, and the gradual expansion of its membership which speaks well of its gradual progress of the bloc toward a multilateral order (Gupta 2006:289). The United States, however, as Kagan points out, seems to be inspired by the Hobbesian view of man in the state of nature who sees the world in an anarchic space. This perfectly justifies the dilemma and uncertainty the United States possesses about multilateralism (Gupta 2006:289).

In the history of Climate Change agreements, the United States has been at the forefront in the process of withdrawing from most of them while resenting the differentiated commitments. It needs to be noted here that the U.S. participation is very substantial in agreements seeking climate regulations and governance, not just owing to its status as a "superpower" but also because the U.S. has the second largest carbon footprint among all countries and is one of the largest emitters of greenhouse gases. However, it has a set record of evading multilateral global regimes which attempt to mitigate and limit emissions among developed countries.

The two trendsetters in the Climate Change Agreement, the 1997 Kyoto Protocol and the 2014 Paris Agreement were both rejected by the U.S. at a point in time, although the latter has enjoyed partisan support. In its analysis, the reasons for the U.S. evasion from these regimes are primarily two — the costs imposed by these agreements and exclusion of developing countries from any commitments on reduction of emissions (Gamble 2010:25). Even though the Kyoto Protocol incorporated the U.S. position in negotiations while the Paris Agreement enjoyed wide support from the Obama Administration, the U.S. attempted to evade both the treaties in concern about its economy and loss of competitiveness toward developing countries exempt from binding commitments. This evasion leads to two grave problems. Firstly, it leads to a total ineffectiveness of the agreement as forestated. Secondly, it leads to the risk of the remainder of the developed countries not complying to the regime in concern for their competitiveness with the U.S. industries as well as fear of a rise in costs of production (Gupta 2006:289). No matter how attractive multilateralism is when it comes to solving global problems, as long as countries like the United States do not see the merit of multilateral institutions, its worth is lost. Perhaps there is a need to accommodate the U.S. interest by multilateralism. But the prospect of this accommodation would mean the abolition of differentiated principles which in itself means abolition of this hybrid multilateralism based on affirmative principles. The United States' hesitations in joining any multilateral agreement is also somewhere justified, as I shall further explore in the section discussing the Free Rider problem. The economic costs imposed by such regimes upon the "responsible powers" would have made proper sense only if the developing world accepted some of the commitments. If the developing world with large CO_2 emitters like India and China continue with their trends while countries like the U.S.A. invest in emission reduction and a green economy, it would be futile to think of a world which would successfully mitigate the risk of climate change. After all, the developing world with its lack of capacities is likely to face the worst risks of climate change and the path of evading commitments should not be the one sought. The proposition here is not to suggest that the developing world is to hold equal responsibility as that of the developed world, but as long as multilateralism and its effectiveness is in question, the hybrid form of multilateralism with all its high principles of affirmative action would never work, at least not in the presence of Hobbesian inspired countries like the United States. A "shared responsibility" implies the developing world to be shareholders in binding commitments too. The drastic effects of climate change must be acted upon by all the states at varying levels lest any mechanism would fail. Contrary to multilateral engagement stands a bilateral regime which is grounded on shared responsibilities and equal commitments. The United States has had a record of engaging in bilateral engagements for the sake of evading multilateral commitments. This reflects how multilateralism has failed to provide a premise of engagement to all the parties (Gupta 2006:289).

Multilateralism and the Theoretical Conflict

The problem with environmental multilateralism is also a theoretical one. According to Andrew Linklater, the Habermasian inspired critical International Relations theory has three main tenets: the normative, the sociological and the praxeological (Linklater 1998). The role of the normative tenet is to promote human emancipation and expose unjust systematic maladies of exclusion while exploring avenues and making institutions more inclusive and culturally sensitive, in order to nurture dialogic communities that give representation to excluded groups (Eckersley 2012:24). The sociological tenet is rooted in mapping the growth and development of communities and identifying the potential to make this development more inclusive. Lastly, the praxeological tenet is rooted in the task to explore the realization of this potential by harnessing moral resources embedded in existing institutions (Eckersley 2012:24). The idea of multilateralism as is the institutional set up of the UNFCCC seems to be inspired by Critical Theory's normative validity based on the ideal of communicative justice. The idea of communicative justice, according to the Habermasian idea, implies that a norm or agreement is valid only when it has received the unforced consent of all those affected after full and free deliberation (Eckersley 2012). Communicative justice, however, is only a theoretical premise which acts as a yardstick on the basis of which agreements can be judged as free or distorted, inclusive or exclusive and therefore more or less legitimate (Eckersley 2012:24).

This critical theorist view of agreements adept to the multilateral process of negotiating environmental treaties poses a challenge in two distinct premises; primarily on critical theory's idea of representation and secondly on critical theory's blindness to outcomes.

The problem with critical theory's idea of communicative justice is that although it demands presence and deliberation by a maximum number of stakeholders, it neither has a theory of representation, since all those affected must ideally be present nor does it set a limit to whom this "all" constitutes. As Eckersley puts it, "All it (critical theory) can do, in relation to any particular regime, is argue that wider and more diverse representation is always better because it will enable a greater range of standpoints and discourses to be aired, provide a better chance of producing decisions that promote generalizable rather than self-serving interests and therefore ensure the autonomy of the many rather than the few" (Eckersley 2012:24). From a theoretical standpoint, promoting a limitless and unrestricted deliberation to advance principles of justice and equality sounds procedurally assuring. However, when this aspect of communicative justice is engrained in institutions like the UNFCCC and the multilateral negotiating process as a whole, it creates a serious problem. Therefore, there is a need to acknowledge that against the principle of ensuring communicative justice, the impact of climate change on stakeholders is relatively disproportionate, while the forces behind climate change too are varying since different countries have a different contribution with respect to their carbon footprints. Hence, it is important to realize that there are degrees of effectiveness when we talk about climate change, and while it is important to ensure communicative justice while complying to principles of deliberative democracy, the problem which lingers around a multilateral climate change agreement is its extensiveness in inclusion. The UNFCCC process is structured around the principles of communicative justice (Eckersley 2012:24) and deliberative democracy, where any agreement developed within its premises must not miss a complete and absolute consensus, without which no outcome can be imagined. This system has plagued the treaty-making process to an extent where we are drifting through the years with no conclusion or outcomes at all. Copenhagen was a classic example of how the extensive participation and consensus rule in multilateral arrangement led to no outcome at all. At Copenhagen, the outcome reached as a part of extensive negotiation between a restricted group of "friends of chair" could have gone a long way in imposing financial regimes which were agreeable even to the United States, however, citing the consensus rule, and more so because of protests from a small group of countries like Sudan, Venezuela, and Bolivia, the document was discarded, ensuring only that the document was "endorsed" (Eckersley 2012:24). When the Conference of Parties met again at Cancun, the proposal to recognize the Copenhagen Accord for its basic mitigation principles and financial pledges was yet again opposed by merely the veto of one country; Bolivia, and thus, all hopes were lost of COP15 to find any success. However, the Mexican chair at COP15 gaveled the veto, proclaiming that one country cannot merely exercise a veto (Eckersley 2012:24). Such episodes at multilateral settings portray the inherent problem of multilateral institutions which lies in extensive deliberation and consensus.

We also ought to note that factors like climate change bear a factor of time where swift action is needed to mitigate effects. Each day lost to negotiations pushes us closer to the effects of climate change for years to an extent where we are already at a point where mitigation of impacts of climate change is not our sole concern anymore. We now also have to think about how to protect the vulnerable countries from a factor which now seems inevitable as we pass through years to a point of no return. Unconstrained dialogues and high principles of justice and deliberative democracy show their limitations when viewed by factor of time. The problem with critical theory and its incorporation in a multilateral approach is that it is blind to time because it is blind to outcomes (Eckersley 2012:28). While a great deal of stress lies in making the procedure just, the multilateral perspective turns blind when it comes to prioritizing outcomes over procedures, as it prioritizes procedural justice over substantive justice. It is futile to ensure more inclusive dialogues while failing to reach agreements within the stipulated time which defies the purpose of dialogues itself. Multilateralism, with its ongoing obsession with principles of inclusion and consensus, can lead to an ultimate doom if it does not prioritize outcomes over procedure, even if it means a compromise in the core structures of multilateralism.

The Free Rider Menace

Game theory suggests three potential areas of agreements in the international arena. The first kind is where reaching an agreement is easy, owing to beneficial and strong incentives for parties; the second class of agreement bears a medium difficulty with principles based on mutual reciprocity like those involved in international trade agreements (Nordhaus 2015). The third class of agreements involve public goods and are confronted with hard problems. Among so many treaties and mechanisms dealing with public goods like those on nuclear non-proliferation, cyber warfare, laws of the sea, etc., climate change is one of the crucial ones (Nordhaus 2015:1339). While engaging in agreements regarding public goods, countries face what is known as the "prisoner's dilemma". As Nordhaus puts it, "The prisoner's dilemma occurs in a strategic situation in which the actors have incentives to make themselves better off at the expense of other parties. The result is that all parties are worse off". This problem is most evident in climate change regimes where countries tend to shrug off their commitments owing mainly to lack of incentives offered by these treaties. As forestated, the impacts of climate change are disproportionate and so what incentive does a developing country have for reducing industrial production by reducing emissions to mitigate the impacts of climate change in a small developing country? This lack of incentive coupled with the hesitancy of the developing world to take on any commitments at all leads to the "free-rider" (Nordhaus 2015:1339). Precisely, the root of hesitancy toward multilateral climate regimes has stemmed from its favorability toward free riding, where the commitments are mostly voluntary on the developed world while absent from the developing world. If multilateralism is to work efficiently and successfully, it cannot aim to achieve that unless it addresses the free rider issue as well as places a good incentive structure upon its parties. Obsessed with the principle of equality, however, this still remains a distant dream.

The Third World Skepticism

Underlying all these issues is the endless skepticism from the Third World to join multilateral alliances even if they involve "common but differentiated responsibilities". Sisir Gupta's speculation that the structure of the world politics has barely had any space for the third world (Gupta 1969:54) holds false today since one of the struggles of the multilateral setting has been directed to taking the developing nations into confidence, but what stands true of his speculation is, "the foreign policy preoccupations of many of the Third World states are more varied than one seeking structural changes in world politics" (Gupta 1969:54). This trend has been very evident over time. The accord at Copenhagen had a representation of the developed world as well as the developing world with the United States and BASIC countries at the negotiating table. Yet, it was a majority of the developing world which opposed the outcome at Copenhagen citing reasons of undemocratic and unrepresentative procedures. Kenyan President Mwai Kibaki most staunchly expressed his dissatisfaction as he exclaimed, "Why am I here? To rubber stamp what? We will not accept this" (Kamau *et al.* 2018).

Alarmingly, the developing world was also a staunch critic of the UNEP's call toward a "green economy" based on renewable energy sources as a shift from "brown economy" based on fossil fuels. The Bolivarian Alliance for the Peoples of Our America (ALBA), a group of Latin American Countries suspected this call to be an attempt by the developed world to capitalize the environmental arena nature is seen as "capital" for producing "tradable environmental goods and services that should then be valued in monetary terms and assigned a price so that they can be used to obtain profits". Citing their ethical principles, this group also rejected any attempt to strategize or monetize toward any such attempt. Some developing countries also thought this attempt to be a plan for European countries to push for products and bring advantage to the green technology producers (Kamau *et al.* 2018).

Until the mistrust between the developing and developed countries persists, no amount of multilateral institutionalism or hybrid multilateralism can converge these consensus-based agreements to act toward climate change. In fact, a trust mechanism should be ingrained in an agreement to make sure that no amount of misunderstandings persist between the developed and the developed world. A compromise from both sides is needed which can never be reached without a sense of reliability and trust between both the worlds.

The Knowledge Gap: A Global North and South Divide

This section analyses the conditions that underpin the knowledge or capacity gap between the Global North and Global South, consequently characterizing what the gap entails. Further, we explore the impact of the knowledge gap on global environmental governance, and essentially, how it ultimately leads to the inevitable failure of the multilateral system.

A growing corpus of literature points out the ever-widening gap of knowledge or capacity between the Global North and the Global South — or between the developed and developing countries. For instance, upwards of 95% of papers in the field of the environmental sciences, biodiversity, ecology, and related fields belonged to authors coming from institutions or affiliations to OECD member countries or developed nations. On the contrary, non-OECD member countries, home to more than 80% of the world's population, went on to publish a mere 5% of the research (Karlsson *et al.* 2007:668).

The reasons for the gap in knowledge or capacity transgress broader avenues of research and are grounded in many factors. Significant factors include GDPPC (GDP per capita), which enables substantial investments in Research and Development (R&D), and fuels tertiary education sectors and Education, ensuring a just supply of able scientists and researchers and access to the English language. Other factors of essence comprise Environmental Governance, creating awareness about the need for funding environmental research; and finally, expenditure on R&D, which assures availability of resources that sustain research and development of theories by the governments.

A lack of these factors (or multiple, at times) creates circumstances wherein there is a total absence of research from authors belonging to the Global South. It consequently deprives their scientific community of considerable intellectual capital, which is pertinent in the development of science reflecting the conditions of the Global South.

An acknowledgment of the historical inequalities between the Global North and the Global South, or rather, the Colonizers and the Colonized, is essential to understand the implications of the knowledge gap. It is of essence to note the historical hierarchies and the access to resources between these two worlds. Postcolonial studies reveal how the knowledge institutions of the Global North shape the thinking, marginalization, and material of people belonging to the Global South. It implies that the apparatus of Northern science has strategically made its way to Southern institutions and universities. It inevitably leads knowledge and information to reproduce in contexts alien to its origin.

A far-fetched implication of the knowledge gap prevails on policymaking in countries that belong to the Global South. Two contradicting forces often trouble the policymakers: Firstly, scientists from the Global South often face difficulties as their work gets less recognition and the empiricism of the knowledge produced by scientists in these countries. Consequently, the lack of access hampers their international credibility as scientists. Secondly, it is the tendency of policymakers to rely on research and knowledge from the Global North or the United Nations. These two factors lead to the disregard of research that is culturally centric to a given environment and subsequently creates space to use research pushed forward by authors from the Global North, ultimately leading to the invisibility of Southern environmental issues in governance (Vadrot 2020).

A comprehensive revaluation of how the knowledge institutions of the Global North exert their authority and dominance over those from the Global South is of paramount importance because there is a far-reaching consequence of the same, as most of the empirical data/research seek to suit the needs of the context it represents. The needs and demands of countries from the Global South differ significantly from that of the Global North. Therefore, it creates dissonance between the policy and the conditions on ground.

An acknowledgment of the knowledge gap is crucial to enable intervention. It would be critical in ensuring a smooth flow of knowledge from the Global South. It would lead to the visibility of contexts and issues absent from the scientific community of the Global South.

Even though multilateral institutions seek to work through the "common but differentiated responsibilities" understanding, as long as the knowledge divide exists, it is impossible to project the Global South in an equal and shared setting. Any policy which seeks to impose commitments upon the Global South will ultimately originate in a country in the Global North, which possesses higher knowledge than the countries adopting such policy instruments.

How then do multilateral regimes expect to judge a country's inclusion or exclusion from a commitment regime when they lack an understanding of the knowledge divide between the Global North and the South? The Global South tends to disagree with a binding commitment, owing to their realization of this lack of capacity and knowledge gap while the Global North tends to enforce these commitments on the Global South owing to their superior knowledge and a lack of realization of the divide that exists between them. It leads to a conflict that would render any multilateral arrangement futile.

Any institution which intends to work toward a long-term future, must engrain within its structure the idea of this knowledge divide and must arrange itself in a way to address it, unlike multilateralism which has so far only contributed to deepening it. Multilateralism has only led to the deepening of the knowledge gap because multilateral agreements tend to ignore the existence of the knowledge divide. It is, thus, fundamental for institutions working on long-term growth and stability to realize this gap and facilitate a system that focuses on reducing this divide.

Exploring Alternatives

As with all things that are complex, breaking them down into more simple and manageable portions is efficient to pave the way for big solutions. It is in this light that minilateralism presents itself as a viable alternative to multilateralism. Its presence in the Asia-Pacific has already started gaining notice with the revival of Quadrilateral Security Dialogue (QUAD) due to the growing dominance of China. With minilateralism comes a focused approach, since unlike multilateralism, only the states which have a direct involvement in a particular matter have a say in it. With the removal of unnecessary diplomatic dialogue thanks to a reduction in participants, the true stakeholders can stand to benefit with a much faster approach to solutions, as opposed to the painstaking duration of multilateral efforts which are way bigger and less targeted in their nature.

Minilateralism is a pathway to focus on the atomic nature of humongous problems. The Pareto Principle, which is more so known as the "80/20 rule" in pop culture can explain the reasoning behind the minilateral approach. When 80% of the impact in a large battle can be created by just involving 20% of the people, it does not make sense to drop this certainty in favor of routine processes that have brought no real advantage to the Climate Change Fight. In the very same light lies the creation of climate clubs, a concept which has found its support echoed by various notable scholars. The main benefit of such climate clubs is the fact that states can join with concretely put-together goals and incentives, alongside targets that a small group of states is more likely to achieve compared to multilateral efforts. The very nature of the word "club" represents an exclusivity and originality that is likely to drive states to work together in a string of cohesiveness and cooperation, something which has rarely been achievable in large forums. It is also a much-needed measure to combat free-riding which was seen in reference to the Kyoto Protocol (Nordhaus 2015:1339). Substantial progress has been achieved already with some hybrid forms of minilateralism.

Examples of such hybrid forms are included in informal institutions such as Friends of Chair in the COP, which was responsible for the Copenhagen Accord, BASIC group, etc. In institutionalized forms, minilateralism forms have been seen in alliances such as the Alliance of Small Island States (AOSIS), African Group (AG), Least Developed Countries (LDC) to name a few. However, the problem with these institutions has been that instead of contributing to a shared agenda and a common goal, they have only played a part in presenting constrained agendas. There is a need to move toward minilateral institutions which contribute to common agendas, which is all ever important thanks to the Paris Agreement, and those which are confluencing spaces for the developed and the developing world. The problem with minilateral institutions lies in the menace of constraining their membership: the "number problem"; who all do we include when the number of stakeholders in an agreement has to be low by nature? (Falkner 2016:87). Unlike multilateralism, minilateralism seeks to move toward a solution by limiting the number of stakeholders. Here, keeping in account the time factor, prioritization is given to substantive justice of critical theory, rather than procedural justice and deliberative democracy. It seems to solve the problem when it comes to scrutinizing an immediate and long-term agreement. However, the question arises when we think about who the primary stakeholders are going to be in a minilateral setting. In this regard, Moisés Naim has argued that the solution to the multilateral number problem with respect to climate change is to involve the smallest number of countries possible, that is, "the smallest possible number of countries needed to have the largest possible impact on solving a particular problem. Think of this as minilateralism's magic number" (Eckersley 2012:32). This would "break the untenable gridlock" which the multilateral setup at the UNFCCC is riddled with. This approach has been successful in the economic sphere, however, there have also been attempts to bring this process into climate change negotiations. The problem with this approach lies in scrutinizing the "magic number" and deciding as to which stakeholders can be involved in such negotiations and what the criteria is for that decision. David Victor furthered the use of this approach after disappointment from Copenhagen by arguing that "more progress will come from small groups of pivotal nations rather than global forums" (Eckersley 2012:33). This claim itself creates further problems since it has a tendency to make the conclusions of any such negotiations self-serving in nature. Only involving the biggest contributors to climate change and excluding the victims of it would not just offend the principle of communicative justice but would also remove an important source of advocacy for action (Eckersley 2012:33). Therefore, according to Robyn Eckersley, this would "reduce the quality of the dialogue and eliminate the answerability of the major emitters to the most vulnerable parties during the crucial negotiation phase" (Eckersley 2012:33). To avoid the exclusionary nature of minilateralism, scholars have also suggested pursuing an "inclusive minilateralism", which would have a proportionate confluence of the most capable, most responsible and most vulnerable actors. In the UNFCCC process, this would mean prioritizing the "common but differentiated responsibilities" principle in a minilateral setting. Different scholars have given different names to this minilateral solution and have suggested different models for determining the "magic number".

Robyn Eckersley has an interesting solution to the problem mentioned above. As forestated, the impacts of climate change differ in degree and so do the actors contributing to the process of them changing. If minilateral institutions are to provide the solution, then such institutions must be constituted of representation with capability, responsibility, and vulnerability. It also becomes important to ensure that these minilateral institutions are integrated in the UNFCCC process.

Conclusion

To conclude, the need for an agreement toward a joint effort against climate change is urgent and multilateralism with its principles and structural flaws has consistently failed to address these flaws. The urgent need, therefore, is either to reform the flaws in the multilateral system to move toward a quicker solution or to find a place for an alternative set up within the aegis of the current institution. Moving either way might cost more time but it would lead to a quick solution once the institutions are fixed, without which we will continue to drift toward a point from where there is no return and no solution to seek. It is time for the world community to decide and reform the broken institutions of multilateralism immediately.

References

Eckersley, R. (2012). Moving forward in the climate negotiations: Multilateralism or minilateralism? *Global Environmental Politics*, *12*(2), 24–42.

Falkner, R. (2016). A minilateral solution for global climate change? On bargaining efficiency, club benefits, and international legitimacy. *Perspectives on Politics*, 14(1), 87–101.

Gamble, A. (2010). *The Politics of Deadlocks. Deadlocks in Multilateral Negotiations — Causes and Solutions*. New York: Cambridge University Press, pp. 25–47.

Gupta, J. (2006). Environmental multilateralism under challenge? In Newman, E., Thakur, R., and Tirman. J. *Multilateralism under Challenge? Power, International Order, and Structural Change*. New York: United Nations University Press, pp. 289–307.

Gupta, S. (1969). The third world and the great powers. *The Annals of the American Academy of Political and Social Science*, *386*(1), 54–63.

Kamau, M., Chasek, P., and O'Connor, D. (2018). *Transforming Multilateral Diplomacy: The Inside Story of the Sustainable Development Goals*. London: Routledge.

Karlsson, S., Srebotnjak, T., and Gonzales, P. (2007). Understanding the North–South knowledge divide and its implications for policy: A quantitative analysis of the generation of scientific knowledge in the environmental sciences. *Environmental Science & Policy*, *10*(7–8), 668–684.

Linklater, A. (1998). *The Transformation of Political Community: Ethical Foundations of the Post-Westphalian Era*. South Carolina: University of South Carolina Press.

Nordhaus, W. (2015). Climate clubs: Overcoming free-riding in international climate policy. *American Economic Review*, *105*(4), 1339–1370.

Ruggie, J. G. (1992). Multilateralism: The anatomy of an institution. *International Organization*, *46*(3), 561–598.

UNFCCC (1992). *Framework Convention on Climate Change*. New York: United Nations.

Vadrot, A. B. (2020). Multilateralism as a "site" of struggle over environmental knowledge: The North-South divide. *Critical Policy Studies*, *14*(2), 233–245.

Section III

Climate Change Narratives with a Focus on India

Chapter 10

International Climate Governance: Indian Perspectives

Chaitra C.*

Abstract

Climate change is currently placed on every national and international agenda and is also shaping the strategies of many businesses worldwide, as international cooperation is indispensable to meet the collective goods problem. Climate policy has evolved as a subset of the foreign policy of major and small, vulnerable island nations. New Delhi has repeatedly repositioned itself from normative to empirical approaches seeking equitable distribution of the carbon space while underlining the narrative of historical responsibilities. Given its emissions profile, economic growth, and leadership role in the developing world, India has been capitalizing on the avenues of climate change-related market mechanisms like clean development mechanisms (CDMs) and co-benefits from the adaptation and mitigation efforts. Over the years, India's climate change policies having emerged as a by-product of its energy and economic policies, in accordance with international agreements and protocols.

India's ambitious initiatives like the International Solar Alliance and Coalition for Disaster Resilient Infrastructure, India's Nationally

---

*Dr. Chaitra C, Assistant Professor in Political Science, Govt. First Grade College, C S Pura, Karnataka.

Determined Contributions (NDCs) to United Nations Framework Convention on Climate Change (UNFCCC), and ratification of Paris Treaty reflect India's engagement on a global scale. At Glasgow, India committed to net-zero emissions by 2070. With a continued emphasis on sustainable development and not compromising its national interests, a multilateral approach will be the way ahead for India. India should continue to be a champion of equity and draw its own low carbon path trajectory, thereby showcasing its responsibilities toward global commons and its actions, forcing the developed countries to introspect their contributions to protecting the planet (PIB 2021b). India's approach to climate governance mirrors balancing international commitments and domestic developmental imperatives. India has a rich tradition of not exploiting nature, and its indigenous knowledge and sustainable lifestyles are already being endorsed on climate negotiation fora (Parampara 2015).

Keywords: Sustainable Development, Climate Governance, Equity

Introduction

The history of climate change negotiations involves power relationships and the intricate linkages to economics, politics, security, and science. Climate change is a contemporary non-traditional threat affecting all nations — some may be more vulnerable, while some others less susceptible. Every individual directly or indirectly contributes to climate change and in turn is a victim of its effects. State and non-state actors need to engage constructively to handle the menace of climate change. Given its demographic profile, biodiversity, energy/emissions profile, economic profile, and rising international stature, India's presence in the international climate governance arena is increasingly felt. However, between 1870 and 2017, India's emissions have been a mere 4% of the global share.[1]

India needs to balance its development agenda as well as adhere to international commitments, while also keeping a check on its carbon footprint (Jayaram 2018). However, it is also true that as compared to India's population, energy resources consumption is meagre. In 2016–2017, India's per capita energy consumption was 22,351 MJ, just one-third of the world average (Economic Survey 2016–2017). While India has overperformed its commitments, it continues to uphold its national interests in

---

[1]CAIT, *Cumulative CO_2 Emissions Including LULUCF in tCO2e 2014.* CAIT Climate Data Explorer, 2018.

the negotiations. As a responsible country, India has been a part of the UNFCCC, the Conference of Parties (COPs) to UNFCCC, most importantly, a signatory of the Paris Treaty, 2015 and asserting its carbon space as a developing country. While India needs to address its domestic concerns, it also needs to adopt a multilateral approach in consonance with countries of the North as well as the South.

Under the Paris Agreement, India had committed to reduce greenhouse gas emission intensity of its GDP by 33–35% below 2005 levels by 2030 and has achieved emission intensity of GDP reduced by 21% below 2005 levels by 2014. Further, it has committed that about 40% of its electricity requirements would be based on non-fossil fuels by 2030 and about 35% of power capacity was based on non-fossil fuels by March 2018 (PIB 2015). India had also pledged to create an additional carbon sink of 2.5–3 billion tons of carbon dioxide equivalent through additional forest and tree cover by 2030. Indian Pavilion at the Conference of Parties (COP) had showcased indigenous knowledge and sustainable lifestyles which promises to enrich its soft power as well. The Parampara and Samanvay catalog that India unveiled at Paris were significant in this regard. Environmental ethos being integral to its civilizational legacy has undergirded India's increasing participation in international climate change governance (PIB 2015).

Simultaneously, India needs to address socio-economic development needs of 1.3 billion people, such as access to electricity to all, and also cater to industrial and infrastructural developments. Being the third largest emitter of carbon dioxide, India needs to move away from rhetorical advocacy appeal as a developing nation. However, India should revive third-world solidarity like in the early years of independence with the Non-Aligned Movement. This third-world cooperation will enhance India's credibility as a global leader for the LDCs and the small island countries (Michaelowa and Michaelowa 2012).

While it needs to insist on technology transfer and financial assistance from the developed world, it also needs to partner with them to fast forward its own initiatives. It is also true that India has often been under-committing and over-performing in terms of its voluntary pledges at various environmental conferences, which is also true with its own domestic renewable energy targets. India also needs to strengthen its relations with the like-minded nations of the South like China, Brazil, and South Africa such as in the BASIC grouping and with oil-producing nations that shares similar concerns and consider legally binding emission reduction commitments as a threat to future economic development potentials.

Climate change continues to be on the agenda of BRICS platform during consecutive years. Establishing friendly relations with the oil-producing countries is essential for India's future energy security, given its low oil-reservoirs and increasing energy demands. Furthermore, if India becomes a member of Nuclear Suppliers' Group, it could probably more effectively implement the Paris Treaty commitments (*Times of India* 2021) and the climate goal under the SDGs framework. The UNSC permanent seat could add to India's increased presence in the international climate governance mechanism.

India's forest cover stood at 701,495 square kilometers in 2015 and raised to 708,273 square kilometers in 2017.[2] However, India is not in a position as yet to dismantle its fossil fuel dependency. The G7 "nature compact", 30by30 pledge formulated by the U.S. has been endorsed by India. India has joined the Adaptation Action Coalition but is not a part of Carbon Offsetting and Reduction Scheme for International Aviation (CORSIA). With 17.8% of the world's population, India is responsible for a meager 3.2% of cumulative emissions.[3] From a climate denial mode to being an agenda-setter, India's role in international climate governance is conspicuous.

India's Climate Diplomacy Narratives

India's normative approach to international climate governance includes the principle of equity and Common but Differentiated Responsibilities (CBDR) and equal per capita access to resources (Agarwal and Narain 1991). In the COP 24 at Katowice, India asserted that "we all agree that the Paris Agreement is non-negotiable. Therefore, the delicate balance reached between developed and developing countries must be retained, and the principles such as equity and CBDR-RC must be given its due".[4]

Several narratives in the climate governance of India have emerged over time, as identified by Navroz Dubash. The first type comprises the *Growth-First Stonewallers* (Dubash 2009), who consider the international climate negotiations as a geopolitical strategy of the developed countries to contain emerging economies like India. International negotiations

---

[2] India's National REDD+ Strategy, MOEFCCC, 2018.
[3] Second Biennial Update Report to the UNFCCC, December 31, 2018.
[4] India Country Statement for COP 24, UNFCCC.

enforce stonewall commitments for India. Thus, they prefer a weak climate regime that would not hinder India's growth pattern and would not require sacrificing the carbon space allocated to it rightfully. They prioritize India's higher growth rate in tune with China's environmentally unconstrained growth spurt for about two decades. Equity is a potential device to keep the developed countries on the opposite axis, and about equity within the nation, they justify it to be more of a strategy than a principle. This position was often a reflection of the Government of India in the past.

The second type, *Progressive Realists*, opines prospective climate impacts as a serious threat to India. They are deeply cynical about the international process, which does not address historical responsibility or advancing equity. They consider India's growth is often used as a pretext for the inaction of the developed world. Further, the notion of North hiding behind India has been a hurdle in the climate negotiations. Therefore, they propose a per capita-based burden-sharing architecture as the justifiable way forward to ensure equity. In the domestic realm, they press for sustainable development-based developmental programs and benefit from subsequent climate co-benefits. In other words, India should do its part but without formally or legally linking it to the international process. The ideas are evident, for instance, in the National Action Plan on Climate Change of India and the insistence that India would subject actions externally supported to the MRV process while the rest would remain outside the scrutiny of the international process. Also, the business communities in India have increasingly considered Clean Development Mechanisms (CDMs) as a business opportunity. This is a reflection of the perspective of environmental and development civil society groups in India.

Finally, the *Progressive Internationalists* who share ideas in common with progressive realists and strongly feel that India can and should make a difference to global negotiation dynamics by explicitly aligning Indian interests with a strong global climate regime. Since the poor are most vulnerable to climate impacts, an ineffective agreement would further perpetuate inequality. Thus, climate issues should address equity as well as effectiveness. Lastly, they significantly emphasize the potential economic opportunities by being the first movers in developing low-carbon technologies and other mitigation efforts. Only a small percentage of people subscribe to this view in India. With a strong record of low carbon initiatives to address development, India has benefited from the CDM process. At the same time, it continues to emphasize the historical responsibilities of industrialized countries. Government, civil society, academia, and business groups have been joining hands for a low carbon future. Instead of putting forward quantitative targets and inflating the business-as-usual scenario, the bottom-up approach could complement the emerging economies' growth trajectory.

Thus, India has a legacy of value-loaded arguments based on its colonial past and current developmental imperatives and the fact that it cannot be equated with the developed world or be forced to sacrifice its carbon space given the developmental needs of its vast population and low levels of poverty.

India's International Presence

India has been a part of almost all the environment-related agreements under the aegis of the UN, most importantly, UNFCCC (1992). India has been a member of Stockholm Conference (1972), Convention on Biological Diversity (1992), UN Convention to Combat Desertification (1994), Rotterdam Convention (1998), Cartegena Protocol (2000), REDD and REDD+ (2005), Vienna Convention (1985), Montreal Protocol (1987), Kyoto Protocol to UNFCCC (1997), Ramsar Convention (1971), CITES (1973), Nagoya Protocol (2010) Minamata Convention (2013), and most recently Paris Agreement (2015) and Kigali Amendment (2016). India has been a member of the IUCN, WWF, IPCC, UNEP, WMO, and several other transnational organizations. Over the years, India's climate change policies having emerged as a by-product of its energy and economic policies, in accordance with international agreements and protocols (Aniruddh 2017).

India remains committed to global efforts as seen in collaborative, ambitious initiatives like the International Solar Alliance (ISA) (Economic Survey 2017–2018). It is a conclave of 124 countries placed completely or partly between the Tropic of Cancer and Tropic of Capricorn aimed at reducing dependence on fossil fuels and utilizing solar energy efficiently. ISA has recently launched two new initiatives — a "World Solar Bank" and "One Sun One World One Grid Initiative". The proposed World Solar Bank would cater to the need for dedicated financing window for solar energy projects across the members of ISA. It is expected to provide low-cost financing at favorable terms for solar energy projects as well as engage in co-financing with other multilateral/bilateral development financial institutions. The "One Sun One World One Grid" vision aims to create an interconnected green grid that will enable solar energy generation in regions with high potential and facilitate its evacuation to demand centers. ISA's progress in solar rooftop program has been equally noteworthy, with a demand of more than 1 GW aggregated from member countries. ISA has diversified its programmatic focus onto health sector, cold storage chains for agriculture and vaccines, and other innovative applications of solar energy.

Alongside, ISA, India launched two initiatives — Coalition for Disaster Resilient Infrastructure (CDRI) to serve as a platform to generate and exchange knowledge related to climate and disaster-resilient infrastructure and Leadership Group for Industry Transition (LGIT), jointly with Sweden, to serve as a platform for interaction between the government and private sector to cooperate in accelerating low carbon growth and technology innovation. The CDRI functions as an inclusive multistakeholder platform led and managed by national governments, where knowledge is generated and exchanged on different aspects of disaster resilience of infrastructure. As of December 2020, 19 countries and 4 multilateral organizations have become members of the coalition. The CDRI is co-chaired by India and the United Kingdom (UK). It intends to "promote resilience of new and existing infrastructure systems to climate and disaster risks". It focuses on developing resilience in ecological infrastructure, socio-economic infrastructure with emphasis on health, education, transport, telecommunication, energy, and water. The LGIT launched in collaboration with the governments of India and Sweden, in September 2019, explores as to how energy-intensive industry can and must progress on low-carbon pathways, aiming at net-zero carbon emissions by 2050.

Furthermore, India's Nationally Determined Contributions (NDCs) to United Nations Framework Convention on Climate Change (UNFCCC) and ratification of Paris Treaty reflect India's engagement with climate change initiatives on a global scale. Under the Paris Agreement, India had committed to reduce greenhouse gas emission intensity of its GDP by 33–35% below 2005 levels by 2030 and has achieved emission intensity of GDP reduced by 21% below 2005 levels by 2014. Further, it has committed that about 40% of its electricity requirements would be based on non-fossil fuels by 2030 and about 35% of power capacity was based on non-fossil fuels by March 2018. India had also pledged to create an additional carbon sink of 2.5–3 billion tons of carbon dioxide equivalent through additional forest and tree cover by 2030 (PIB 2015). Further, India is on its track to successfully decoupling its economic growth from GHG emissions. As per the second BUR submitted to UNFCCC in 2018, India's emission intensity of GDP reduced by 21% in 2014 over the level of 2005. To ensure the use of cleaner automobile fuel, India has also leapfrogged from BS-IV to BS-VI emission norms on April 1, 2020, earlier than the initial date for adoption in 2024 (Economic Survey 2020–2021).

Furthermore, India's participation and assertion at non-UNFCCC processes were evident. India has not been a signatory to CORSIA again on the basis of the argument of differentiated responsibilities (Sinha 2019). The agreement requires aviation companies to report their carbon emissions accurately to the national authorities and should compensate their emissions through carbon-offsetting which is not a cost-effective process for the developing countries and the LDCs. India has been a signatory to Montreal Protocol and was also a part of the Kigali Amendment to the Protocol (2016) which meant to phase-out Hydro fluorocarbons (HFCs) emission and agreed to a timeline to reduce the use of HFCs by 80–85% of their baselines over the next several decades. According to International Energy Agency (IEA), the share of space cooling in peak electricity load is projected to rise sharply in India from 10% of the current period to 45% in 2050.

Mission Innovation (MI) launched on November 30, 2015 during UNFCCC COP-21, is a global initiative to accelerate widespread public and private clean energy innovation for an effective long-term global response toward climate challenge. The aim is to provide affordable and reliable energy for everyone and promote economic growth, which is critical for energy security. India, along with 22 other countries and the EU is a member of all seven MI challenges for clean energy development and is a co-lead in three challenges on smart grid, off-grid, and sustainable bio-fuel.

India is a member of the Workshop on GHG Inventories in Asia (WGIA), a network of Asian countries on GHG inventory, initiated by Government of Japan to assist countries to improve the quality of their inventories by promoting the exchange of information and experience obtained in the region. India participates in annual WGIA sessions, and the 16th WGIA was organized in India in 2018. Bilaterally, India has entered into Memorandum of Understanding with a number of countries to exchange and strengthen expertise on climate change mitigation and adaptation matters during the reporting period of this BUR.

India was a signatory to the Hyogo Framework for Action (2005–2015) and also to its successor Sendai Framework for Disaster Risk Reduction (2015–2030). India had embraced Millennium Development Goals (MDGs) (2000–2015) and its successor Sustainable Development Goals (SDGs) (2015–2030). Sendai Framework, SDGs, and Paris Treaty should be carried out in coordination with each other to sustain a better future.

India asserted on clean energy, clean fuel, clean water at the Adaptation Action Coalition formed at Climate Adaptation Summit, 2021. India asserted that it would not only reduce environmental degradation but also reverse it. India has pledged to restore 26 million hectares of degraded land by 2030. Thus, India has been demonstrating that it will do more than its fair share of responsibilities (Shoko 2021).

India and COP 26

The COP 26 in Glasgow during October 31–November 12, 2021 was once again a witness to enhanced participation of India. India unveiled its five-fold strategy (Panchamrit Strategy) to the world (PIB 2021a). It included an increase in its non-fossil energy capacity to 500 GW by 2030. Adhering to its earlier commitment, India has increased its renewable energy targets to 450 GW by 2030. Secondly, India committed to meet about 50% of its energy requirements from renewable energy by 2030. Thirdly, India assured to reduce the total projected carbon emissions by one billion tons from the present times until 2030. Fourthly, India has committed to reduce the carbon intensity of its economy by 45%. India has achieved 25% of emission intensity reduction of GDP between 2005 and 2016 and is on the path to achieve more than 40% by 2030. Lastly, India as part of its long-term strategy assured to achieve Net Zero by 2070. Indian Prime Minister gave the slogan of One Life One World with LIFE implying *Lifestyle for Environment Today* which again was an indication to adopt an environment-friendly lifestyle (Das and Chaturvedi 2021). India also launched the E-Amrit Portal on electric vehicles during the summit.

However, India was one of the last major carbon emitters to declare a net-zero date. China, Russia, and Saudi Arabia have committed for the same by 2060. While the U.S., U.K., and EU have committed to a 2050 deadline, Germany and Sweden have committed to 2045 target. India has not been a part of the 130-countries' coalition aiming at ending and reversing deforestation by 2030 but at the same time led the Infrastructure for Resilient Island States to help small island states.

Further, at Glasgow India preferred to use "phase down" of coal instead of "phase out" which was criticized by many countries (PTI 2021). But this does not actually imply that India is diverting from its international commitments. About two-thirds of the needs of Indian population is fulfilled by coal and therein lies the inevitability of such a usage. At the same time, India's ambitious commitments at Glasgow provides "massive investment opportunities in the renewables, EV ecosystem, ethanol blending, energy efficiency, carbon capture technologies",[5] points the rating agency ICRA. There are at the same time several ambiguities in India's commitments at Glasgow (Powell Lydia *et al.* 2021). Furthermore, since the COP 26 finalized the rules of carbon trading, India shall be in a position to sell more than a million carbon credits from previous years and also be able to create a domestic market for carbon trading as pointed out by Ulka Kelkar of World Resources Institute India (PTI 2021). Bhupendra Yadav, India's Environment Minister reiterated that the current climate crisis has been precipitated by unsustainable lifestyles and wasteful consumption patterns in the developed countries (Nandi 2021).

The Way Forward

The third-world cooperation will enhance India's credibility as a global leader for the LDCs and the small island countries. While India needs to insist on technology transfer and financial assistance from the developed world, it also needs to partner with them to fast forward its own initiatives. It is also true that India has often been under-committing and over-performing in terms of its voluntary pledges at various environmental conferences, which is also true with its own domestic renewable energy targets. India also needs to strengthen its relations with the like-minded nations of the South like China, Brazil, and South Africa such as in the BASIC grouping and with oil-producing nations that shares similar concerns and consider legally binding emission reduction commitments as a threat to future economic development potentials. Climate change continues to be on the agenda of BRICS platform during consecutive years.

---

[5] ICRA, *India's COP 26 Commitments to Help with New Green Technologies*, January 5, 2022.

Establishing friendly relations with the oil-producing countries is essential for India's future energy security, given its low oil-reservoirs and increasing energy demands. Furthermore, if India becomes a member of Nuclear Suppliers' Group, it could probably more effectively implement the Paris Treaty commitments and the climate goal under the SDGs framework. The UNSC permanent seat could add to India's increased presence in the international climate governance mechanism. The glaring challenge for India will remain the phasing out of coal. It is estimated that for India to achieve carbon neutrality by 2070, coal for power generation needs to drop by 99% by 2060.

The emphasis on net zero by 2050, under the aegis of the UN, to adhere to 2°C target and China's announcement of achieving net-zero by 2060 places further burden on India. Amid the COVID pandemic, green recovery and circular economy arrangements have got an impetus. Thus, with a continued emphasis on sustainable development and not compromising its national interests, a multilateral approach will be the way ahead for India. India should continue to be a champion of equity and draw its own low carbon path trajectory, thereby showcasing its responsibilities toward global commons and its actions, forcing the developed countries to introspect their contributions to protecting the planet. India's approach to climate governance mirrors balancing international commitments and domestic developmental imperatives. India engages in greening by stealth by abiding by its voluntary commitments, thereby increasing its respect and bargaining space in the international realm. It should uphold its citizens' well-being and prosperity and ensure energy security and prestige in the international realm (Vihma 2011). India has a rich tradition of not exploiting nature, and its indigenous knowledge and sustainable lifestyles are already being endorsed on climate negotiation fora.

Indian Prime Minister has also earlier announced the goal of attaining "energy independence" by 2047 (Dixit 2021). Shailly Kedia, TERI, recently opined, "India's normative and entrepreneurial leadership is key to achieving the goals of climate stabilization". Several initiatives have been showcasing India's commitments. India has signed the five-year green strategic partnership with Denmark. However, India has not joined the Global Methane Pledge floated by the European Union.

In conclusion, India's domestic climate policy and international strategies and priorities are defined by the SDGs and NDCs in the current times. A developing country like India, has been adhering to international commitments by being an active part of the negotiations, signatory of environment related treaties and protocols, and been devising effective strategies to facilitate climate change adaptation, mitigation, resilience, and disaster management. Meanwhile, India has also taken advantage of the CDM market and is investing increasingly in renewable energy sources. The national and subnational policies and programs related to climate change are intertwined with aspects of finance and energy portfolio. The academia, civil society, market along with the government apparatus have been facilitating India in achieving its ambitious climate actions.

References

Agarwal, A. and Narain, S. (1991). *Carbon Emissions in an Unequal World.* New Delhi: Centre for Science and Environment.

Das, P. and Chaturvedi, V. India at COP 26 and beyond. *ISAS Insights*, December 23.

Dixit, R. (2021). Climate change summit: What will India's position be? *The Week*, October 24.

Dubash, N. (2009). *Towards a Progressive Indian and Global Climate Politics.* New Delhi: Centre for Policy Research Climate Initiative, pp. 9–12.

Economic Survey (2016–2017). *Climate Change Sustainable Development and Energy*, Vol. 2, Chap. 5, p. 121

Economic Survey (2017–2018). *Sustainable Development, Energy and Climate Change*, Vol. 2, Chap. 5, p. 73.

Economic Survey (2020–2021). *Sustainable Development and Climate Change*, Vol. 2, Chap. 6, pp. 208–228.

Jayaram, D. (2018). From spoiler to bridging nation: The reshaping of India's climate diplomacy. *Cairn International*, *109*(1), 181–190.

Michaelowa, K. and Michaelowa, A. (2012). India as an emerging power in international climate negotiations. *Climate Policy*, *12*(5), 55–59.

Mohan, A. (2017). *From Rio to Paris: India in Global Climate Politics.* New Delhi: ORF.

Nandi, J. (2021). India leads negotiations as COP 26 deal is done. *Hindustan Times*, November 15.

Nodo, S. (2021). Delivering climate action: The road ahead for India after COP 26. *Down to earth*, December 16.

PIB (2015). *MOEFCCC*, December 29.

PIB (2021a). *National Statement by Shri Narendra Modi at COP 26 Summit in Glasgow.* November 1.

PIB (2021b). *Shri Bhupendra Yadav Delivers Statement on Behalf of BASIC Countries at COP 26.* November 1.

Powell, L., Sati, A., and Tomar, V. (2021). *India's COP 26 Pledges: Ambitious, But Ambiguous.* New Delhi: Observer Research Foundation.

PTI (2021). Climate experts support India's stand on "phase down" of coal at COP 26, November 14.

Sinha, S. (2019). India seeks changes in ICAO's calculation for offsetting aviation greenhouses gas emissions. *The Times of India*, September 29.

Times of India (2021). *NSG Pitch: For Climate and Strategic Goals, West Must Help India Expand Nuclear Power Infrastructure*, November 3.

Vihma, A. (2011). India and the global climate governance: Between principles and pragmatism. *The Journal of Environment & Development, 1*, 69–94.

https://doi.org/10.1142/9789811263750_0011

Chapter 11

Policies as Instruments in Promoting Sustainable Development: Limiting the Climate Change Issues in India

Sheeraz Ahmad Alaie*

Abstract

Climate change is the most critical issue threatening the whole world. As a most populous, tropical developing country, India has a bigger challenge to cope with the consequences of climate change than other countries. However, climate change is a global phenomenon but consequences are both local and global, so the role of policies both at global and country level is very crucial. The Indian Climate Change policy is featured under both regional and global dimensions with National Action Plan on Climate Change (NAPCC) adopted in 2008 and India's Intended Nationally Determined Commitments (INDC) 2015, respectively. The chapter tries to assess the role of such policies in promoting sustainable development and will ascertain the challenges and opportunities within such policy perspectives. It will try to decipher whether there exists any substantive institutional framework for policy which offers coherence and consistency as to how the government should cope

---

*Postdoctoral Fellow, DST-Center for Policy Research, Panjab University Chandigarh, 160014. Email: elahisheeraz10@gmail.com.

with the long-term political challenges of climate change. The critical subject to ponder upon is the characterization of India within the international climate landscape "as a minor contributor to past emissions, but a significant contributor to future emissions, albeit not on a per capita basis" (Dubash *et al.* 2018). India's national climate policy urgently needs a coherent vision for tackling climate change, that should be reflected in framing of legislation and policy documents addressing carbon emission-related sectors and aligned with several federal levels, and in the strategy of appropriate institutional frameworks to achieve the climate policy objectives of adaptation and mitigation.

Keywords: Climate Change, Policies, Sustainable Development, Carbon Emissions, Institutional Frameworks

Introduction

Climate change resulting through global warming is observed through rising temperatures, shifting weather patterns, rising sea levels, and increasing storm intensity. The global warming itself is caused due to emission of greenhouse gases (GHG) from human activities like, fossil fuel usage, agricultural activities, and deforestation. This issue under the sustainable development goals has been labeled as a priority at the global level to minimize the emission of GHGs. Since 2000, most developed countries have implemented policies at the country level to address the significant issues of climate change and related problems, however marking it as a global crisis that has attracted efforts at the international levels (Sachs 2015:84). OECD states that large reductions are achievable at relatively low costs, if proper policies are formulated and implemented, including strong use of market-based instruments globally to develop a standard global prize for GHG emissions. To reinforce the achievements, better integration of climate change objectives is required in specific policy areas such as transport, energy, infrastructure, agriculture or forestry, and other parameters to speed technological innovation and diffusion. The climate change issues have led policymakers from various countries to devise ways of limiting emissions of GHGs. The policy instrument to be employed has to keep in balance both progress in industrialization and subsequently control of GHGs emissions. Stabilizing concentrations of GHGs in the atmosphere at a relatively rigorous level can be accomplished at costs less than 1/10th of 1% of gross domestic product (GDP) growth per annum, or less than 3% loss in GDP by 2030. These figures are indicated in the Fourth Assessment Report Intergovernmental Panel on Climate Change (IPCC), supported by recent OECD estimates. Such relatively low-cost estimates of GHG emission reduction undertake widespread use of economically efficient market-based policy instruments, such as carbon taxes and trading emissions, and broad participation in mitigation efforts across the world. One of the significant tools under mitigation is the climate funding or financing from developed countries to developing countries. To cite an example, the developed countries pledged to mobilize $100 billion in international climate finance per year by 2020, at the Copenhagen climate summit in 2009 to help developing countries cope with the impact of climate change and to promote transition toward low-carbon development. The promise was reiterated in the Paris Agreement of 2015. Such assistance to cope with the climate change issues is commonly referred to as international climate finance (Gupta *et al.* 2014:1238).

India occupies a significant position in climate change-related problems. Its CO_2 emissions rose slower in 2016–2019 than in 2011–2015 but was above the world average of 0.7% (Koshy 2021). In comparison to the U.S. and China, India's carbon emissions are very low. In 2018, China emitted 10 billion tons of CO_2, the U.S. emitted 5.41 billion tons of CO_2, while India produced only 2.65 tons in the same year. Moreover, no country was performing well enough to meet the 2015 Paris Agreement goals. The United States, with a rank of 61, was the worst performer. India, continued to remain in the top 10, and scored 63.98 points out of 100. It achieved good ratings on all CCPI indicators except for "renewable energy", where it was categorized as a "medium" performance country. Indian stance is that, India did not cause climate change as compared to the developed countries do. Hence, it reiterates the climate change issues with adaptation and mitigation strategies under schematic policy within national and global perspectives. However, climate change is a global phenomenon with far-reaching consequences that are both local and global, so the role of policies both at global and country level is very crucial. The Indian Climate Change policy is featured under both regional and global dimensions with National Action Plan on Climate Change (NAPCC) adopted in 2008 and India's Intended Nationally Determined Commitments (INDC) 2015, respectively. The assessment of these existing domestic and foreign Indian climate policies is very voluminous and complex as they comprise various missions and agreements. Here in this chapter, we are assessing the impact of such polices in a holistic manner, without specifying any particular mission or agreement. The parameters used to understand the impact of climate policies are scoped within the domain of low carbon emission, resources conservation, climate issues awareness, and fossil replacing green technologies in the Indian context. The study is qualitative in nature based on the secondary data related to climate change polices and agreements of India. The section below high-lights the understanding of how policy measures are taken as instruments to find solutions for socio-economic problems.

Policies as Instruments

A policy instrument is a linkage between policy formulation and policy implementation which acts as governing tools to achieve policy targets of managing resources but adjusted to political, social, economic and admin-istrative concerns (Ali 2012:99). Selecting a particular policy instrument with problem-specific nature attains policy success for controlling anthro-pogenic influence on resources. There exists a large number of policy instruments systematically applied to control the scale of resource flow to mitigate ups and downs in policy. Multiple instrument selection is required if policy implementation involves multiple levels of government or inter-sectoral communities. For climate change solutions, Schmalensee (1996) advocated the creation of durable institutions and frameworks rather than a particular policy program. Pertinent to global climate change, two categories of policy instruments are implemented by most of the countries. These two categories are domestic national policy instruments and international policy instruments (bilateral, multilateral, or global). National policy actions seek to enable specific nations to achieve specific goals or targets related to climate change.

The international policy instruments require two or more than two countries to regulate climate change at the global level. It is stated by Nordhaus and Yang (1996:748) that, individual nations determine energy and environmental policy, so any grand framework to lessen global warming must be translated into national measures. However, it is also advocated that successful policies to address the real global environmen-tal problem will need the adoption of international agreements. The policy instruments applied by the EU in areas of nature conservation or its related issues, are legislative and regulatory instruments, economic and fiscal instruments, agreement-based or cooperative instruments, information and communication instruments and knowledge and innovation instruments (Bocher 2012; Lascoumes and Le Galès 2007). Other studies have used four principal policy instruments in environmental policy, such as, environmental effectiveness, cost-effectiveness, distributional considerations, and institutional feasibility. In this chapter, both approaches are used to understand the impact of such policies in combating the climate change.

National Action Plan for Climate Change (NAPCC): An Outlook

NAPCC has been proposed as a strategic cross-sectoral instrument to control the climate change causes by limiting the emission of GHGs and conserving resources. The NAPCC aims at creating awareness among the public, government agencies, scientists, industry, and the community as a whole against the threats posed by climate change and the strategies proposed at national level to counter these changes. Under this plan, eight sub-plans were introduced to promote sustainable development through limiting the GHG emissions. It was built around the establishment of eight national "missions" aimed at integrating mitigation and adaptation aspects of climate change into national policies across a range of sectors. Some of these missions had specific focus and targets, such as a Solar Mission aimed at enabling 20,000 MW of solar power by 2022 (MNRE 2009). Others, such as a National Water Mission, have broader and more diffuse objectives including water conservation, creation of a database, and promotion of basin level integrated water management (MWR 2009). Other missions focus on energy efficiency, agriculture, Himalayan ecosystems, sustainable agriculture, sustainable habitat, a "green India" mission focused on the forest sector, and a strategic knowledge mission. Among all these missions, the solar mission got much priority and earlier implementation.

The objective of the Water Mission is to ensure integrated water resource management to conserve water, to distribute it equally, and minimize wastages both across and within states. The mission will consider provisions of National Water Policy and develop a framework to optimize water use by increasing its usage efficiency through regulatory mechanisms with different pricing and entitlements. This mission ensures waste water recycling as well as adoption of new and appropriate technologies to provide adequate water facilities for coastal cities. The storage of water both above and below the ground are planned to be enhanced by technologies and infrastructures with efficient management structures. The water mission is planned to get institutionalized by respective ministries and organized through inter-sectoral groups including Ministry of Finance and Think Tank agencies. Despite the national mission on water, ground water depletion has been experienced in various areas especially where this water is used for irrigation of large fields. The mission needs more robustness in planning and awareness among people to add value to this precious natural resource.

A strong strategic knowledge system, one of the national missions for climate change actions, is significant for identifying, planning, formulating, and implementing policy-driven engagements, while maintaining the essential economic growth rate. Such a strategic knowledge-based system for informing and supporting climate perceptive actions will be vital to address a number of determined objectives. The mission should discourse climate science with area specific modeling; an assessment of various technology scenarios and alternatives for complying with feasible national objectives; leveraging global cooperation and initiatives for selection and development of new innovations and technologies for adaptation and mitigation; and hence bridging knowledge gaps. It is important that the vitality of the knowledge enterprise directing and managing climate change issues is sustained through human and institutional capacity-building. These measures are essential for designing policy responses and implementation set approaches at the national level and for inputs to negotiate the international for a by respective designated departments.

The Mission for Green India, as one of the eight national missions under the NAPCC, recognizes that climate change experiences will seriously affect and alter the type, quality, and distribution of biological resources of the country and the associated sustainable livelihoods of the people. Mission for a Green India recognizes the impacts that the forestry sector has on environmental betterment though climate mitigation, water security, food security, biodiversity conservation, and sustainable livelihood security of forest reliant communities. To promote this mission the objectives established are likened to increased forest or tree cover on 5 m ha of forest or non-forest lands and improved quality of forest cover on another 5 m ha (a total of 10 m ha) of land; improve ecosystem services including biodiversity, hydrological services, and carbon sequestration as a result of treatment of 10 m ha; increased forest-based sustainable livelihood income of about 3 million households living in and around the forests and enhanced annual CO_2 sequestration by 50–60 million tons in the year 2020.

The Mission (National Mission for Sustaining the Himalayan Ecosystem) needs to convey better understanding of the coupling between the climate factors and Himalayan ecosystem and provide inputs for promoting Himalayan sustainable development, also addressing the protection of a fragile ecosystem. It requires the joint effort of scientists, climatologists, glaciologists, and other experts. Exchange of knowledge, experience, and information with the South Asian countries and countries sharing the Himalayan ecology will also be helpful. There is a need to create an observational and monitoring network for the Himalayan environment to evaluate freshwater resources and health line of the ecosystem. The mission makes efforts to address some pressing issues concerning Himalayan Glaciers and the associated hydrological consequences, biodiversity conservation and resource protection, wild life conservation and protection, traditional knowledge societies, and their livelihood and planning for sustaining of the Himalayan Ecosystem.

The mission on agricultural sustainable development, titled officially as the National Mission for Sustainable Agriculture (NMSA), brought up the strategies and programs of actions (POA) that aim at promoting sustainable agriculture through a series of adaptation measures focusing on 10 key dimensions encompassing Indian agriculture namely; "Improved Crop Seeds, Livestock and Fish Cultures", "Water Use Efficiency", "Pest Management", "Improved Farm Practices", "Nutrient Management", "Agricultural Insurance", "Credit Support", "Markets", "Access to Information", and "Livelihood Diversification". NMSA will cater to key dimensions of "Water use efficiency", "Nutrient Management", and "Livelihood Diversification" through adoption of sustainable development pathway by progressively shifting to environmentally friendly technologies, adoption of energy efficient equipment, conservation of natural resources, integrated farming, etc. The national policy for climate change has framed the respective institutions to gear up sectoral level approaches to combat carbon emissions. The resource management at the multi-institutional level may prove to be productive in limiting climate change.

India's Intended Nationally Determined Commitments (INDC) 2015

All countries worldwide adopted a historic global climate agreement at the United Nations Framework Convention on Climate Change (UNFCCC) Conference of the Parties (COP21) in December 2015 in Paris. In anticipation of this convention, countries publicly charted what post-2020 climate actions they projected to take under the new international climate-related agreement, known as Intended Nationally Determined Contributions (INDCs) for each specific country. The climate change-related actions communicated in these INDCs largely determine whether the world achieves the broad-term goals of the Paris Agreement to imply efforts to limit the increase to 1.5°C, hold the increase in global average temperature to well below 2°C, and to attain net zero emissions in the second half of the present century. INDC addresses national policy settings where countries determine their potential contributions in the context of their national priorities, capabilities, and circumstances within an international framework under the Paris Agreement driving collective action toward a zero-carbon emission, climate-resilient future. The INDCs generate a constructive feedback circle between national and international policy and decision-making on climate change.

The pledge under India's INDC includes a 33–35% emissions intensity reduction level by 2030, a renewable energy capacity pledge which is lower than India's domestic target, and the forest cover target. These interpreted as "par" targets require little effort beyond existing policies (Dubash and Khosla 2015:12). The MOEFCC set up an implementation committee that did not function properly and failed to prepare NDC implementation plans (Arora 2020). In 2020, the MOEFCC again constituted a cross-ministerial "apex committee" of bureaucrats to oversee NDC progress and revision, and to regulate carbon markets (Ministry of Environment, Forests and Climate Change, 2020). But the lack of a high-level strategic oversight role lead to contradictory efforts, such as the promotion of renewable energy simultaneously that has characterized India's approach, most recently as part of its COVID recovery strategy (Sengupta 2019). The Indian position in international climate negotiations and domestic climate policy debates is expressed in troubling binaries: economic development versus climate change mitigation, centralized command-and-control environmental governance regimes versus decentralized adaptive governance mechanisms, transitioning to renewable energy versus carbon sequestering through forests, and so on. The characterization of India in the international climate landscape is, "as a minor contributor to past emissions, but a significant contributor to future emissions, albeit not on a per capita basis" (Dubash *et al.* 2018).

The quantification of goals in India's NDCs is threefold: reducing the emission-intensity of its gross domestic product (GDP) by 33–35% by 2030; achieving 40% cumulative electric power installed capacity from non-fossil fuel-based energy sources by 2030; and third, establishing an additional carbon sink of 2.5–3 billion tons of CO_2 equivalent by 2030 through additional forest and tree cover (Nandi 2021). India has significantly improved its ranking in the Climate Change Performance Index 2019 by rising up three places higher to rank 11 (Behl 2018). The government's pledge to properly strengthen its renewable energy capacity, coupled with economic factors, such as falling renewable energy prices (Dubash *et al.* 2018), suggests that India may meet its NDCs target of achieving 40% electric power installed capacity from non-fossil fuel based resources well ahead of schedule (Goswami 2018). Similarly, India is also expected to achieve its quantified NDCs goal of reducing GHG emission intensity (Sethi 2018) to achieve sustainability.

The non-quantified objectives in India's NDCs comprise putting forward and propagating a healthy and sustainable way of ease-of-living based on traditions and values of conservation and moderation; adopting a climate friendly and cleaner path than the one followed by others at corresponding levels of economic development; mobilizing domestic and new additional funds from developed countries to implement mitigation and adaptation actions; better adapting to climate change by improving investments in development programs in sectors vulnerable to climate change; and building capacities and establishing a domestic framework and international architecture for quick diffusion of cutting-edge climate technology in India and for joint collaborative research and development for such future technologies (GoI 2016).

INDC sets a clear signal for green and clean energy, i.e., achieving its target of 40% non-fossil-based energy by 2030 while generating 200 GW on new power capacity by 2030. The emission intensity in India declined by 18% approximately between 1990 and 2005 and the target commits to reach 33–35% by 2030. India's INDC signifies the importance of prominently restoring forest cover to act as a sink for carbon along with supporting livelihoods. This additional carbon sink creation for about 3 billion tons of CO_2 would require average annual carbon sequestration to increase at least by 14% in the future years. Keeping in view the vulnerability issues toward climate change, heavy focus on adaptation and resilience in India's INDC exist, highlighting the current initiatives in sensitive sectors, like water, agriculture, health, and more. It spends about 3% of its GDP on adaptation. INDC noted that increased investment in these activities will need additional support both from national and international funding agencies, estimating a requirement of $206 billion for 2015–2030 period (Mitra *et al.* 2015). The current INDC for period 2021–2030 in response to decisions 1/CP.19 and 1/CP.20 makes submission of points like, sustainable way of living with values of moderation and conservation; cleaner climate/eco-friendly path for economic development; reducing emissions intensity of its GDP by 33–35% by 2030; and mobilizing domestic and additional funds of mitigation.

Climate Change Causes: India's Perspective

The United Nations Framework Convention on Climate Change (UNFCCC) required the developed countries to undertake promising domestic action plans to regulate global temperature below 2°C. It also implored developed countries to make contribution in new and additional international assistance for developing countries based on general but differentiated responsibilities and respective capabilities to address climate change. China, India, and EU represent 40% of global carbon emissions, set to achieve more than their agreement in the Paris Agreement 2015 first round. As mentioned earlier, China and the United States are major emitters (see Figure 1).

India's approach to climate change politics is based on the fact that its per capita carbon emissions are around a third of the global average. However, it was listed as fourth largest emitter in year 2018 reporting annual emissions that has been growing at just under 4% between 1994 and 2014 (Friedlingstein *et al.* 2020). Its carbon emission trajectory is consequently subjected to frequent international pressure. Indian climate politics has long been based on the premise that increases in per capita energy use are needed to bridge India's development deficit (Dubash *et al.* 2018). It is reported, that 73% of emissions are derived from the energy sector and 42% from power generation (Government of India 2018). Energy has always been a significant topic for India's domestic political economy, from national level electoral competition about the

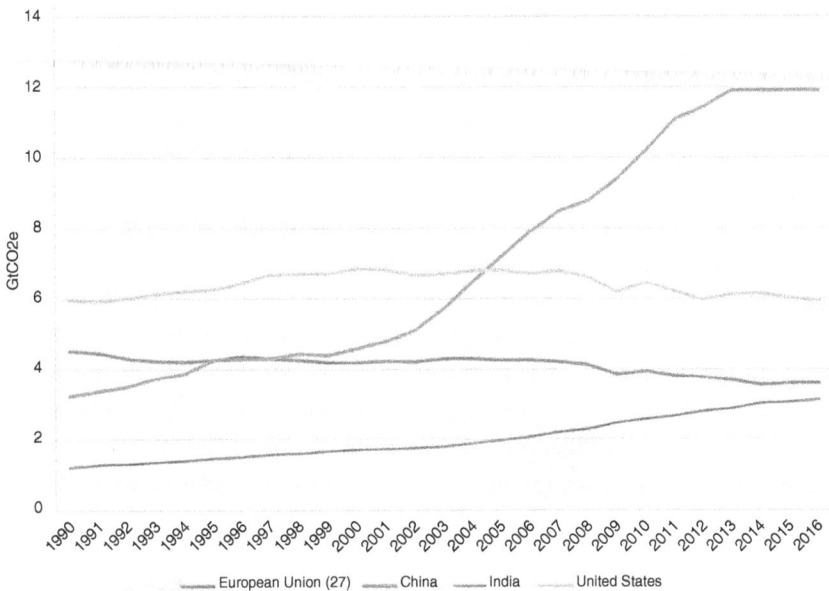

Figure 1: India's annual emissions in comparative context (excludes land use change and forestry emissions).

Source: World Resources Institute 2020.

affordability and reliability of electricity in state politics (Dubash *et al.* 2018) to politics around cooking fuel access (Varma and Bhaskar 2018). Mohapatra and Giri recommend appropriate environmental regulations that can substantially stimulate innovations to increase energy efficiency and thereby reduce carbon dioxide (CO_2) emissions (Mohapatra and Giri 2021). Mitigation policy has evidently been overshadowed by developmental politics. In the late 2010s, greater international pressure on developing countries and shifts in India's domestic narrative featured establishing new institutions. Since climate change continued under the radar of domestic politics, these institutions were shaped more by bureaucratic politics and routines than by domestic political institutions.

In the fiscal year 2019/2020 due to pandemic lockdowns, carbon dioxide (CO_2) emissions in India fell by an estimated 30 million tons of CO_2 ($MtCO_2$). This was the first time emissions had fallen in the country in four decades, with a drop of three $MtCO_2$ last occurring in 1982. Although factors such as renewable energy growth and economic

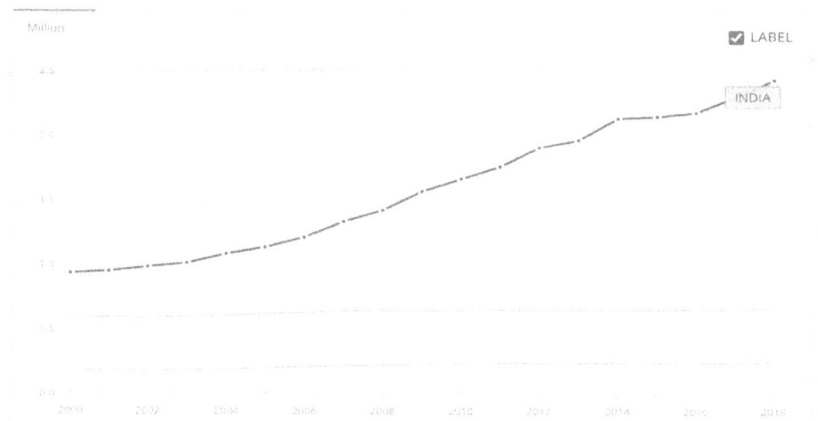

Figure 2:   CO$_2$ emissions (kt) India.
Source: World Bank, 2021.

slowdown affected these figures, measures, such as lockdowns, implemented to slow the spread of the novel coronavirus (COVID-19) had a big impact in early 2020. According to the source, March 2020 experienced an estimated 15% drop in emissions when compared to the previous year (Tiseo 2020). Before the pandemic year, it is observed that a sharp rise of carbon emission is reported for India for year 2000–2018. The trend of carbon emission in India from the year 2000–2018 can be viewed in Figure 2.

Figure 2 shows that despite instituting domestic and international climatic policies, India portrays a trend of increasing global carbon emissions. India is world's fourth largest carbon emitter, responsible for 6% of GHG emissions. However, a little reduction has been reported due to the introduction of solar technologies. The World Bank report shows that there is an increasing rather than a decreasing trend, depicting the weakness of policies. There are several ongoing programs for raising of funds for climate change mitigation. One of the programs like Green Growth and Equity Fund advocates investment in low carbon and climate resilient platforms across the energy value chain, comprising the renewable energy production, energy clean, and efficient technologies, low carbon transport and resource conservation. Another funding program is Enhancing Climate Resilience of India's Coastal Communities. India's coastline is expected to be among the most affected by climate change. Climate change impacts such as extreme weather events and sea level rise are exacerbated by urbanization, overfishing, and poorly planned coastal development. This means that approximately 250 million people (14% of the country's population) who live within 50 kilometers of coastal areas are particularly vulnerable to climate change. One more program namely, "Line of Credit for Solar Rooftop Segment for Commercial, Industrial and Residential Housing Sectors" is mandated to enhance the harnessing of solar energy. In its Nationally Determined Contribution (NDC), the Government of India has stated its ambition to achieve 40% cumulative electric power capacity from non-fossil fuel-based energy resources by 2030 — with a target of 40 GW of rooftop solar power by 2022. The program will increase access to long-term and affordable financing for the construction of 250 MW of rooftop solar capacity in India and thereby reduce emissions by 5.2 million tons of CO_2 equivalent over 20 years.

Conclusion

India has conventionally approached climate change as a significant diplomatic issue, by insisting that the highly industrialized countries take a lead in reducing the carbon emissions and subsequently providing support for developing countries through finance and technology. To combat climate change at the national level, a large number of missions under one holistic plan depicts the lacking of stringent road mapping to achieve goals. It shows a divergent nature of the plan with multiple pathways and features of each mission. These include a lack of clarity on emissions intensity in the base year (2005) and target year (2030), as well as the scope and coverage of the intensity target and the methodologies for measuring it. The overall policy for climate change action in India is two-fold, *viz.* national and international with one goal to promote sustainable development. The domestic policies seek adaptation to climate change and international policies focus on mitigation of the adverse effects of climate change. The impact of policies on climate change in India has led India to adopt newer technologies which can replace old high carbon emission technologies. The sustainable mission has rigorously been identified in solar-based technology sector; however, in other areas like ground water depletion and air pollution, rising carbon emissions are still on rise. The information related to climate change issues is significant for monitoring achievements toward India's target and for understanding how it

214 S. A. Alaie

contributes to the international goal of limiting temperature rise to 1.5°C. India has put forward a well-balanced climate plan that — alongside its renewable energy goals — will generate transformational changes. These actions are likely being proposed alongside a prominent development agenda. Although some implementation challenges remain, the INDC makes it clear that India along with its allies is working tirelessly toward a strong international climate agreement in future.

References

Ali, M. (2012). *Sustainability Assessment: Context of Resource and Environmental Policy.* Cambridge: Academic Press.

Arora, S. (2020). Confederation of Indian Industry and member of a sub-committee, January 8.

Behl, M. (2018, December 11). India's ranking improves, but coal-based power plants threaten clean energy bid, *Times of India.* https://timesofindia. indiatimes.com/city/nagpur/indias-ranking-improves/but-coal-based-power-plants-threaten-clean-energy/bid/articleshow/67032602.cms.

Bocher, M. (2012). A theoretical framework for explaining the choice of instruments in environmental policy. *Forest Policy and Economics*, 16, 14–22.

Dubash, N. K. and Khosla, R. (2015). Neither brake nor accelerator: Assessing India's climate contribution. *Economic and Political Weekly*, 50, 10–14.

Dubash, N. K., Khosla, R., Rao, N. D., and Bhardwaj, A. (2018). India's energy and emissions future: An interpretive analysis of model scenarios. *Environmental Research Letters*, 13(7), 074018.

Friedlingstein, P., O'sullivan, M., Jones, M. W., Andrew, R. M., Hauck, J., Olsen, A., and Zaehle, S. (2020). *Global carbon budget 2020. Earth System Science Data*, 12(4), 3269–3340.

Goswami, U. (2018). India to achieve climate goals before schedule: Environment Minister Harsh Vardhan. *Economic Times*, December 3.

Gupta *et al.* (2014). Cross-cutting investment and finance issues. In Edenhofer, O., Pichs-Madruga, R., Sokona, Y., Farahani, E., Kadner, S., Seyboth, K., Adler, A., Baum, I., Brunner, S., Eickemeier, P., Kriemann, B., Savolainen, J., Schlomer, S., von Stechow, C., Zwickel, T., and Minx, J. C., eds., *Climate Change 2014: Mitigation of Climate Change. Contribution of Working Group III to the Fifth Assessment Report of the Inter-Governmental Panel on Climate Change.* Cambridge: Cambridge University Press.

Kosjy, J. (2021, March 01). India's percentage CO_2 emissions rose faster than the world average. *The Hindu.* https://www.thehindu.com/news/national/indias-percentage-co2-emissionsrose-faster-than-the-world-average/article 33965283.ece.

Lascoumes, P. and Le Galès, P. (2007). Introduction: Understanding public policy through its instruments — From the nature of instruments to the sociology of public policy instrumentation. *Governance, 20*(1), 1–21.

Ministry of Environment, Forests and Climate Change (2020). Climate Change Report.

Ministry of New and Renewable Energy (MNRE) (2009). *Jawaharlal Nehru National Solar Mission: Towards Building Solar India.* New Delhi: Government of India.

Ministry of Water Resources (MWR) (2009). *National Water Mission under National Action Plan on Climate Change: Comprehensive Mission Document,* Vol. 1, p. 61. New Delhi: Government of India.

Mitra, A., Damassa, T., Fransen, T., Stolle, F., and Mogelgaard, K. (2015). *5 Key Takeaways from India's New Climate Plan (INDC).* Washington, D.C.: World Resources Institute.

Mohapatra, G. and Giri, A. K. (2021). The economic growth–income inequality–poverty nexus and its impact on environmental degradation: Empirical evidence from India. *Millennial Asia.* https://doi.org/10.1177/097639 96211023878.

Nandi, J. (2021, August 16). India only G20 nation to meet climate goals. *Hindustan Times.* https://www.hindustantimes.com/environment/india-only-g20-nation-tomeet-climate-goals-101629061426571.html.

Nordhaus, W. D. and Yang, Z. (1996). A regional dynamic general-equilibrium model of alternative climate-change strategies. *The American Economic Review, 86*(4), 741–765.

Sachs, J. D. (2015). *The Age of Sustainable Development.* Columbia: Columbia University Press.

Schmalensee, R. (1996). *Greenhouse Policy Architectures and Institutions.* Cambridge: MIT Joint Program on the Science and Policy of Global Change.

Sengupta, D. (2019). Coal India advances target for one billion tonnes production by two years to 2024. *Economic Times,* November 1.

Sethi, N. (2018). Climate conference faces a crisis as differentiation is diluted in rulebook. *Business Standard,* December 15.

Tiseo, I. (2020, May 25). Annual change in carbon dioxide (CO_2) emissions in India 1982–2020. Emissions, Energy and Environment. *Statista.*

Varma, G. and Bhaskar, U. (2018). How Ujjwala Yojana is emerging as Modi govt.'s MGNREGS ahead of 2019. *Livemint,* May 31.

Chapter 12

The Impact of Fridays for Future Movement in Indian Politics

Sonia Roy*

Abstract

The FFF-India chapter is an interest group comprising primarily young climate activists and volunteers, who seek to use protest demonstrations as a means to direct attention of the government and civil society to the alarming effects of climate change and its impact. According to the website (fridaysforfutureindia.com) — it is an informal and decentralized association set out to achieve a common goal — to protest against global warming, climate change, and perceived government apathy to environmental concerns. FFF interestingly states that this movement "avoids power-structure" and "are beyond politics". Can a movement that is built on the principle of influencing political developments on climate change, remain apolitical in nature? While questioning those in power — (and their policies), how does this organization define it's being beyond politics? What is the future of such a movement? The particular case of the arrest of Disha Ravi, founder and a young climate change activist of Fridays for Future movement in India by the state, received worldwide criticism. Looking into the aggressive stance of the Indian state

---

*Assistant Professor of Political Science, West Bengal Education Service, presently posted at Taki Government College (North 24 Parganas, West Bengal). Email: soniaroy03@gmail.com.

vis-à-vis such climate movements, the chapter would seek to understand the power dynamics at play for these two predominant players — the Indian state on the one hand and the Social Movement Organizations (SMOs) like FFF-India on the other. The hypothesis of this chapter is that despite claiming to be non-political in its objectives, the Fridays for Future movement in India is increasingly acting as a lobby in garnering political influence in Indian politics toward its cause.

Keywords: Fridays for Future, New Social Movements, Greta Thunberg, Disha Ravi, EIA 2020

Introduction

There is a plethora of youth activism through different modes and approaches that have mushroomed around the world today. India remains no exception. Indian youth activism has ranged from the issue of corruption, demand for good governance to that of climate concerns. Online activism, unlike the previous generational modes, has been effective and easily reachable. Educated youths form an integral part of our civil society, many of whom do not yet enjoy the universal franchise. The new social media platforms form an effective means to reach out to the government with their demands. Online activism has given this section of our population a voice to reach out and demand for their rights. Demands range from right to water, clean air, protecting marine life to habitable urbanization (Swarnakar 2019; Luomi 2020). The advantage of these climate activist movements is that they can hold up their views directly to their governments, making them perhaps the largest networked grassroot movement in recent times.

This chapter is divided into five sections. Each section tries to focus on the impact of the Fridays for Future-India movement in the domestic politics, and the larger influence of youth-based activist movements in the country. A social movement for a common cause has the potential to attract like-minded members in the larger civil society. Firstly, they are successfully mainstreaming the cause for climate change mitigation and the role of domestic politics. Secondly, these organizations are actively commandeering the youth of the country, who have been traditionally kept out of activism, for lack of a platform dedicated to their numbers.

In the first section, we trace the origin of the movement to Greta Thunberg, a Swedish national who through her activism has shaped this movement. The second section takes a closer look into the Indian chapter. The focus of this section is to explain the inspirations from the mother movement, and how different parts of India and its youth have successfully captured the national imagination with their climate advocacy.

The third section places the youth movements into a wider context, trying to understand what makes these one of the largest networked grassroot movement in a short time. The members are the youth, many still in schools. The method of propaganda and protest is mainly through online mediums, though physical presence has been successfully witnessed throughout the country since the last couple of years. The fourth section looks exclusively at the government reaction to the *Fridays for Future* movement in India, making a particular note of the hostile tone the government has set toward them. The final section tries to understand the possible future this movement can present in the Indian scenario. Will the movement die out slowly due to the hostility of the government or will its porous borders encourage other similar movements into increasing activism? (Adve 2020).

Greta Thunberg and her *Fridays for Future*

International governmental and non-governmental organizations, like that of the United Nations, tend to take the lead in heralding the nation states toward an organized forum for deliberations regarding climatic change. The local movements (within national boundaries) speaking of environmental concerns often remain unexpressed in the international arena, restraining its vigor within internal borders. Individual expressions as showstoppers are an exception to the norm. Greta Thunberg forms a prominent example to that exception (Kyroglou 2021).

The Swedish environment activist, Greta Thunberg, is credited with a continuously expanding protest movement concerning climate change, actively relating the youth.[1] The Fridays for Future movement creates a platform for the youth in societies, across nations, to organize themselves

---

[1] Newspapers and findings attribute this "Greta Effect" to the fact that increasingly children are now using social media for activism, and people in general seem to be more aware of the ecological cost of airplane rides on the environment. Her activism has inspired people across the world to be more proactive and seek information into climate change action.

into peaceful, yet prominent, expression against the perceived government and corporate apathy and reluctance to protect our environment for future generations (Dunlap and McCright 2008; Dunlap 2014; Batalha and Reynolds 2013).

In August, 2018, at the age of 15 years, Thunberg sat outside the Swedish parliament (*Riksdag*) during school hours, holding up a sign that read, *Skolstrejk för klimatet* (School Strike for Climate). Her demands were for the Swedish parliament to own up to the reduction in carbon emissions, as decreed in the Paris Agreement. The non-violent and politically non-aligned statement has had a visible impact on world leaders and societies across the globe. This campaign, which Thunberg has later framed into the now much popular *Fridays for Future*, is a call toward environment education and advocacy by the youth for a habitable world in the future (Benkenstein *et al.* 2020).

Environment movements and ecological expressions have historically educated people in matters of climate concerns. Issues in climate change and the greenhouse effect remain a subject of discussion in schools today (Conde and Sanchez 2010; Menon 2012). The question then arises, what makes this "Greta Effect"[2] so relevant?

It is this alarming reality of climate change that has encouraged young climate activists across the globe to speak up to their governments, and to the world. Greta Thunberg forms a part of that generation of young climate activists, disillusioned with the empty promises of the political leaders, to save her future.[3] Her journey into activism started at the young age of eight, when she was first educated about climate change in school. Her activism has been inspired by many in her peer group, and she is often hailed as a living inspiration to many of her peers across the globe.

Thunberg has broken barriers of the global north–south divide with her frank anxieties. Her sincere concerns, alarming intensity (visibly outspoken, direct, and often insolent delivery) and her advocacy for climate

---

[2] She was recognized as the Person of the Year by *The Time* in 2019. Her activism has inspired many, and is presently being considered for the 2021 Nobel Peace Prize.

[3] Numerous speeches given on many international platforms show this disdain. Recently, Greta Thunberg spoke at the Youth Climate Summit, 2021 in Milan where she dismissed the political leadership and their concerns on climate change mitigation with her words, "Built back better… bla bla bla". Her speech was both applauded as well as widely critiqued, but makes her position on the global leadership inefficiency on climate concerns more pronounced.

change and mitigation has influenced the Global South with equal vigor. What places her in a unique position is that she situated herself in a platform that is principally "beyond politics" and unquestionably beyond geographical borders. Her use of social media platforms, helps her integrate into a peer generation of equally concerned global citizens. She speaks for her future, united with every youth voice that is concerned for their own. She is unmoved by the global leadership, when they fail to actively address the factual concerns engrossed in their political jargon (Huttunen and Albrecht in Zabern and Tulloch 2021). Speaking at the UN Climate Action Summit in 2019, her words were blunt, speaking her truth to a global audience:

> This is all wrong. I shouldn't be up here. I should be back in school on the other side of the ocean. Yet you all come to us young people for hope. How dare you!
>
> You have stolen my dreams and my childhood with your empty words. And yet I'm one of the lucky ones. People are suffering. People are dying. Entire ecosystems are collapsing. We are in the beginning of a mass extinction, and all you can talk about is money and fairy tales of eternal economic growth. *How dare you!...* (Thunberg, 2019)

While the carbon emissions by the developed countries are the biggest contributor to the environmental emergency that the world is living through, it is those who live in the developing, "third world" countries — the Global South, who are facing the brunt of climatic catastrophes. This frightening truth has acted as a unifying force for young people worldwide, to take a stand for their future, and the right to live on a planet that is not a gangrenous entity.

Fridays for Future — The Indian Chapter: Goals, Limitations, and Potentialities

The *Fridays for Future* instituted its Indian chapter in 2018. The idea behind this movement is to mobilize school students across the world, who voluntarily skip classes every Friday, organizing weekly demonstrations of their demands to their political leaders. Over these Friday demonstrations, these volunteers and activists demand action to prevent climate change, and for the fossil fuel industries to focus on sustainable development measures that would emphasize the creation of renewable and clean (green) energy.

Ever since Thunberg's own social media outreach, this has gained followers and advocacy networks in India, as well as across the globe. They run their own authenticated website, along with social media platforms on Facebook, Twitter, and a LinkedIn page.[4] Their social media presence is regularly updated, making them user-friendly and easily accessible to the present generation of school and college students.

A tread through any of these different social media platforms outline their objectives, goals, and their underlying mission. A newsletter feature regularly updates its observant audience to their present activities and future plans. In order to gain a better comprehension of the *goals* of this Indian chapter, a look at their website provides a good understanding of their viewpoint. Just shy of two years old, this group of committed volunteers has marked their presence in many parts of India, forming their own independent and collaborative regional chapters. The group was first visibly active during the Global Climate Strike in 2019. From 2020, the COVID pandemic has necessitated a change in the protest landscape. Unable to physically unite at locations, the lockdown has given a boost to creative *online* activism, maintaining the urgency and momentum of this program among the youth.

According to the website, accessed through https://www.fridays forfutureindia.com/:

Fridays for Future is a global people's movement for climate justice.
We at *Fridays for Future* India are an inclusive and egalitarian, non-partisan, autonomous and decentralised movement.

Speaking of the perseverance of this people's movement, they also boldly proclaim, "All are welcome. All are needed". The element of inclusivity implies a dynamism which contains within itself the potential of a full-fledged movement for the march against climate change, and demanding sincere social corporate responsibility.

The regional chapters to the website link up to the various movements that are happening at different state-levels in India, while sustaining the

---

[4]For more details, please visit https://www.fridaysforfutureindia.com/.

central pan-India and the global outlook of this concerned people's movement. In order to better understand the movement and its active involvement in Indian climate movement policy advocacy, the Fridays for Future India outlines 10 values that they adhere to. In their own words, these values are as follows:

1. Everyone is welcome. Everyone is needed.
2. We have a mission to ensure that our planet is livable (livable).
3. We work to enable a space of love and acceptance.
4. We enable an environment of regeneration and self-sustenance (sustenance).
5. We strive to dismantle systems of oppression.
6. We actively avoid power structures.
7. We amplify the voices of those at the frontlines of the climate crisis.
8. We are a non-violent movement.
9. We value knowledge sharing.
10. We are beyond politics.

Reiterating the fact that it is a youth-powered movement, the goal of this Indian chapter is to be inclusive. It welcomes citizens from all walks of life, who are concerned about climate change and the severity of ecological imbalance due to carbon emissions. The movement positions itself by amplifying the corporate social responsibility in creating an atmosphere of sustainable development. With a mission to make the planet livable, the outlook remains global even when the focus is on the local scale. A primarily peaceful, non-violent movement — the Fridays for Future India has no unified base. It is a decentralized movement, typical of most people-powered movements, where the activists draw their strength from one another.

The most important factor of this movement is the claim that it is "beyond politics". While on the one hand this movement assumes a position that it "actively avoids power structures", it is often in direct conflict with the power structures that seek to maintain the *status quo*. The most important question that arises is, how can a people powered movement remain politically non-aligned, when the very issues they demand action from is steeped in political actions (or rather, political inactions) by the ruling elite in power. The chapter tries to address this vital query in its examination about the impact of the Fridays for Future Movement in

Indian Politics. From a very precursory glimpse, three things become clear at the very outline:

(1) This particular people's movement is in its nascent stage when compared in terms of various other environmental protest movements that have impacted the Indian political domain.
(2) A decentralized moment consisting of volunteers from the youth in society, its scope and influence remains a matter of apprehensive skepticism.
(3) Being politically non-aligned necessarily means that of being politically neutral. What does the term "beyond politics" mean for its members?

Perhaps the most significant aspect this chapter seeks to realize, or partially understand, is the importance of these new social movements in bringing the agency to the youth of the country. Have the older generations really failed to make their mark where climate change, "eco-anxiety" and its mitigation is concerned? Today, is it the youth who must hold the mantle and make the older generations responsible and accountable to the damage caused to Planet Earth? With all the activism and agreements in the past, have the *adults* really lost the perspective to what is really required, and the urgency with which the present needs to protect the future. This chapter aspires to understand the changing momentum brought forward by the dynamism of these new social movements, and the increasing redundancy of the older generational cause (Rees and Bamberg 2014).

Indian Youth in the Climate Change and Environment Protest Movement — Their Protagonist Capacities and Limitations

These "click activists" or "slacktivists" (as they are known by the older generation of seasoned climate activists in the country), form a new generation of climate change advocates whose outlook is more global and reactionary (Li *et al.* 2021). These new generations have been prominently active in the last three or four years. What remains a matter of special interest about the involvement of the youth in climate change advocacy in the country is that for years before, there just wasn't any.

This is not to say that the Indian environment movements have never involved youth among their numbers. However, the previous generations of climate change activists in the country were usually NGOs and professionals whose aim was to deliberate and raise awareness. The older generations of activism were organized through some social cause or through specified organizational mechanisms — that seeks to influence government policies through formal institutions and lobbying.

This new generation youth climate activism turns this longstanding notion over its head. An essentially decentralized movement, the team structure is flat when it comes to decision-making. Unlike the previous generations, these new social movements that raise issues of national and global importance are much more fluid in its function. Lacking a centralized element, the volunteers come together to organize events, arrange police permits for protests, and organize logistics as and when necessary. The spontaneity of the movement is what primarily separates these youth activists from the earlier ones.

To emphasize this element of difference, Claude Alvares, an older-generation climate activist,[5] finds that the difference between the two generations is actually in the method. His observations point to the largely "online" nature of the new activism. According to him:

> Older activists visit government departments, lobby there, file legal petitions and use a more interpersonal approach, … The government also takes us more seriously because we have a standing for so many years. (Johari and Lalwani 2021)

Nevertheless, these "online" or "click activists" have an advantage over the previous protests' movements. Alvares himself concedes that, the present generation, through social media activism and dissemination of information, have managed to mark a strong presence in the fight against climate change. These tech-savvy teenagers and young in society connect with each other on social issues that they are really concerned about, and can mobilize themselves much more efficiently and effectively. These youth climate activists are truly a people's movement in the correct sense

---

[5] Claude Alphonso Alvares is the Director of the Goa Foundation. An Indian environmentalist, Alvares is a member to many committees like the Supreme Court Monitoring Committee (SCMC) for control of hazardous waste and has been actively involved over the years in Public Interest Litigations on the environmental concerns in the country.

of the term, organizing themselves with an agenda that cuts across borders and into our very survival (Fielding and Hornsey 2016). With attractive sloganeering, and online information, the activism has spread across more quickly and the youth have been participating with real commitments (Dono *et al.* 2010).

Typically, these youth movements carried forward their agenda in the digital platform, through online petitions and letters (e-mails) against these watered-down provisions. Their propaganda and online activism brought them directly in conflict with the government (Dhar 2021). Apart from the Fridays for Future-India, several other youth organizations like the Let India Breathe and There is No Earth B, to name a few, launched simultaneous online campaigns and sent letters to ministries opposing the Act (Dhara 2021). The way the government reprimanded and responded against these activist groups, (comprising mainly school children and the youth in society), has set the tone for the recent power struggle between these youth activist organizations on the one hand, and the institutionalized government reaction on the other (Ferguson *et al.* 2016). The particulars of this power tussle will be discussed in detail in the next section of this chapter.

Speaking of the several limitations of these new age social media motivated climate activism, there remain certain socio-structural constraints. Largely owing to its decentralized nature, there is a lack of hierarchy among this youth climate activist group, including that of the Fridays for Future movement in India. Such a loose formation leaves it open to pressures from the system, and adverse government reactions have resulted in the slowing down of the collective movement. The arrest of youth activists, and the threats of FIRs have brought forward a very serious jolt to these movements, who are at their very nascent stage in India. The societal pressure, from the government as well as from the families have led to a reconsidering among its youth quotient regarding the future of the movement. Fear of their immediate future (scholarships, jobs) have to be weighed carefully against the fear for their future (environment). Many such young activists have taken down social media posts supporting Fridays for Future movement, bringing forward concerns about its future course of action. At present, a glance through its dedicated website reads

"We don't have any events scheduled as of now".

Judged against the structural animosity, both from the government authorities and their support systems (their families), it will be interesting to watch whether these nascent movements, that have so successfully unified the young and the youth in our country, continue to exist and prevail. A lack of funding, no viable protection against legal measures by the government, and the fear-mongering have at present severely limited the momentum of such a vibrant coming-of-age phenomenon the country was witnessing in its youth activism.

The Government Stand *vis-á-vis* Youth Climate Activism in Recent Times, the Particular Case of the Fridays for Future Movement in India

When we speak of the Indian government and their standing where youth activism is concerned, the picture is often tricky to navigate. The most prominent news in recent times was the arrest of one of the founding members of the Fridays for Future movement in India, when she was taken in custody for sharing an edited toolkit on the Twitter platform, retweeted by Greta Thunberg later. Keeping in mind that these are young people, often in schools and colleges, such a drastic step by the Indian state has received worldwide criticism. The disproportionate action by the state against Disha Ravi, a young climate activist from Bengaluru associated with the Indian chapter has brought forward more questions than answers. There were also FIRs registered against three other climate activists, in the same timeline, throwing light on the harsh state position *vis-á-vis* the decentralized climate activism movement in India. The question that then naturally arises is this: What instigates the government to take on such a punitive position on Fridays for Future movement in India? (Bashir 2013).

When we speak of the stand of the government and business houses where climate change mitigation is concerned, we should also keep in mind the concept of corporate social responsibility (CSR).[6] India is an

---

[6] Corporate Social Responsibility entails a legal commitment by the businesses who promise to be more sensitive to the social and environmental concerns of the country. The concept simply underlines the importance of the social responsibility that the corporates have in the betterment of our environment and future.

agrarian country, with a deep historical connection to nature. It is made mandatory for business houses and corporations today to adhere to these stipulated guidelines that will balance the increasing need for development and the sustainability of the Indian ecosystem (Billett 2010). The father of the nation, Mahatma Gandhi himself, recognized the need for his India to maintain and sustain deep connection to nature, and has spoken at length on the need for environment protection and sustainable development in this rush for urbanization and mechanization of the society. His wisdom is contained in these words:

"The Earth has enough resources for our need but not for our greed".

This section will primarily focus on the two particular episodes that are associated with the Fridays for Future and its Indian chapter. The first is that of the Indian governments' reaction to the Fridays for Future-India's opposing stance with regard to EIA 2020. The other incident is that of the more recent arrest of Disha Ravi, and how that has panned out. The reason that these two incidents are being focused upon is simple. The hypothesis that this chapter tries to put forward is that youth climate activist movements are not absolute apolitical in nature as they claim to be, and that they have been successful in influencing a lobby, and opposition support that has influenced the intelligentsia and the civil society to support their prescribed cause.

EIA 2020 and the Fridays for Future-India Chapter

The Environment Impact Assessment Draft, 2020 contributed to the impetus of these youth climate activist collective movements with what was basically seen as its "violator-friendly" nature. This EIA 2020, which is set to replace the earlier 2006 notification, created dissatisfaction among the environment activists in the country owing to its diluted nature where protection of the environment is concerned. According to the United Nations Environment Program (UNEP), EIA is defined as a "tool to identify environmental, social and economic impacts of a project prior to decision-making". This new draft, introduced on March 23, 2020 by the Union Ministry of Environment, Forest and Climate Change (India) is set to bring forward many changes to the existing environmental governance in the country. According to specialists, this *post-facto* grant of clearance would result in unregulated environment degradation and will bring forward a serious challenge to the environment protection regime of the country (Menon 2012; Lohia 2020).

Such *post-facto* clearance will increase the violators, like in the case of the LG Polymers India Pvt. Ltd., who caused a tragic gas leak in its Vishakhapatnam plant due to non-compliance to regulations in place. When we look into the human dimensions, the diluted provisions in the EIA 2020 (draft) would give rise to internally displaced persons (IDP) and increased climate migrants, as their original place of residence will be unfeasible for human habitation, not to mention the ecological imbalance that will be caused by to the violations of the environment protection laws in place. Rapid industrialization, linked to the much-needed economic growth and increased GDP of the country, could come at the cost of the environment. The youth climate activists fear that these provisions would open a floodgate of violations to the well regarded and previously established environment protection and sustainability laws of the country by the corporates and business houses. In brief, the Fridays for Future movement advocated (through the online medium) of the increasing irresponsibility in the CSR and that of the environmental governance in the country.

The subsequent government reaction to the advocacy of the movement caused alarm not only within the country, but has invited international criticism from governments and organizations worldwide. As a consequence, to their protest movement, which were carried out mainly through emails and by protesting peacefully holding placards with their demands, the cyber cell of the Delhi Police brought its full might on their website. This is the main medium through which the Fridays for Future organizes and networks with its members in the country. Added to that, the blocking of their website was carried out under the anti-terror Unlawful Activities Prevention Act (UAPA), which was retracted soon thereafter (citing clerical errors). The act received widespread criticism from all sections of society. Such an unparalleled action by the police forces on a group of youth climate advocates gave rise to serious debates and discussions on what was essentially read-in as the way the state reacted to such an essential democratic movement protected by our fundamental rights in the Indian constitution.

The Arrest of Disha Ravi

A 22-year-old climate activist from Bengaluru, Disha Ravi, was arrested and has spent five days in judicial custody over the controversial toolkit related to the ongoing farmers protest in the country. The police allegedly linked Ravi to a pro-Khalistan movement, with the view of causing political unrest in the country. This caused widespread alarm and reaction from all over the country. According to the Coalition for Environment Justice in India, a statement was brought out in support and signed by "concerned civil society activists and citizens", who saw this action as the state trying to delegitimize the farmers' protests in the country and making a scapegoat of soft targets like the youth in society and the environmental concerns in the country. Many political leaders and prominent personalities in the country came forward in support of Ravi, calling out what was perceived as the absurd state reaction to non-violent movement supporting a just cause. The government has a different side to this. According to the government sources, the editing of the toolkit was an adjunct to the larger cause of organized internal unrest in the society.

Is the Movement Too Nascent? A Possibility of Its Porous Boundaries Encouraging Similar, United Movements in India

What makes the youth climate activism movements, in particular the Fridays for Future movement stand out is its structure. A decentralized, online-based movement, FFF-India has successfully held the national imagination of the youth in society with their firm determination and consistency when it comes to issues of climate change and its mitigation. What is perhaps far more interesting is that this movement, in India and across the world, has a global outlook even though the issues championed by the chapters are often local. The Bangladesh chapter,[7] to take an example, involves issues and concerns associated with the state and its ecology.

Nevertheless, the outlook of this movement retains its global character. There is a constant communication and networking back and forth between the Fridays for Future and its regional parts. This does not mean that the relationship is subordinate. Barring few global programs, these

---

[7]For more details please see, https://fridaysforfuture.org/newsletter/tag/bangladesh/.

regional chapters are completely independent and the structure is flat when it comes to its day-to-day function. It is a voluntary movement, and the volunteers feel for the cause that they champion.

The movement has attracted personalities in support of their cause on a wide scale. Senior Congress leader, Shri Jairam Ramesh, wrote on his verified Twitter account that such an action by the government of arresting Disha Ravi, a founding member with the Indian chapter (Ramesh 2021)

"Completely atrocious! This is unwarranted harassment and intimidation. I express my full solidarity with Disha Ravi".

Many other political leaders and public personalities have also come out in support of the government targeting the Fridays for Future movement. What is interesting to note is that a new-age movement such as this is influencing opinions and making its mark in Indian politics. Twitter outrage on the Indian scene shows the government action against the Fridays for Future is being understood in a larger context. Greta Thunberg (2021) herself remarked on her verified twitter account:

Freedom of speech and the right to peaceful protest and assembly are non-negotiable human rights. These must be a fundamental part of any democracy. #StandWithDishaRavi

This leads to the next of possibilities. Can these collectives lead to organized movement in the future? The presently leaderless, decentralized organization has been successful in uniting people in the name of climate concerns in what is loosely interpreted as a state-corporate nexus in furthering development at the cost of sustainability (Vihma 2011; Johari and Lalwani 2021). Fridays for Future movement, through their online email petitions, Facebook posts, and Twitter/Instagram outreach, have been successful in relaying the message across nations. Intimations and setbacks have only helped its cause, providing it with much needed publicity through media coverage.

Conclusion

Fridays for Future have claimed to be apolitical. What is perhaps truer is that it is politically neutral. Ideologically, it stands for environment protection and climate change mitigation. Its ideology is that of climate activism. One cannot fit this movement in the traditional left-center-right spectrum of political divides. What makes movements and collectives such as this unique in the history of climate activism, is that it moves beyond. In the wake of the Ukrainian crisis, the FFF decided to hold international strikes every Thursday calling for an end to the war in Ukraine (Kamal 2022).

Neutral in its political base, FFF-India has been successful in instigating political actions from both those in power and those who have corporate interest. With their media outreach, it has successfully established responsibility where sustainable development is concerned, and univocally calls out anyone whose interests and acts amount to its deviation. Intimidation aside, this movement has risen to challenge the state on issues such as the Mula-Mutha Riverfront Development Project (RDF), which would result in concretization of the river banks, and impacting the river width, posing threat to the local ecosystem (Sherekar 2022).

The ongoing momentum and increasing dynamism of the movement promises a future that will remain relevant and reflective of the society through its youth-based advocacy. Such a leaderless online movement has perhaps found a new protest base, of being a true people's movement in the real sense of the term.

References

Adve, N. (2021). Coming of age of India's youth climate movement. *The India Forum*, March 24. https://www.theindiaforum.in/article/coming-age-india-s-youth-climate-movement.

Bashir, N. Y. *et al.* (2013). The ironic impact of activists: Negative stereotypes reduce social change influence. *European Journal of Social Psychology*, *43*, 614–626.

Batalha, L. and Reynolds, K. J. (2012), Aspiring to mitigate climate change: Superordinate identity in global climate negotiations. *Political Psychology*, *33*, 743–760.

Benkenstein *et al.* (2020). Youth, civil society organizations and academia, *South African Institute of International Affairs.*

Billett, S. (2010), Dividing climate change: Global warming in the Indian mass media. *Climatic Change.*

"Completely atrocious": Congress leader Jairam Ramesh "expresses full solidarity" with Disha Ravi, held in Greta Thunberg toolkit case (2021). *The Free*

Press Journal, February 14. https://www.freepressjournal.in/india/completely-atrocious-congress-leader-jairam-ramesh-expresses-full-solidarity-with-disha-ravi-held-in-greta-thunberg-toolkit-case.

Conde, M. C. and Sanchez, S. (2010). The school curriculum and environment education: A school environmental audit experience. *International Journal of Environment and Science Education, 5*(4), 477–494.

Dhar, A. (2021). Youth climate action and political transformation in India. *Sciences Po*, June 14. https://www.sciencespo.fr/psia/chair-sustainable-development/2021/06/14/youth-climate-action-and-political-transformations-in-india/.

Dhara, T. (2021). The young environmental groups leading India's new climate activism. *The Caravan*, March 15.

Dono, J. *et al.* (2010). The relationship between environmental activism, pro-environmental behaviour and social identity. *Journal of Environmental Psychology, 30*, 178–186.

Dunlap, R. E. (2014). Clarifying anti-reflexivity: Conservative opposition to impact science and scientific evidence. *Environmental Research Letters, 9*, 021001.

Dunlap, R. E. and McCright, A. M. (2008). A widening gap: Republican and democratic views on climate change. *Environment: Science and Policy Sustainable Development, 50*, 26–35.

Fielding, K. S. and Hornsey, M. J. (2016). A social identity analysis of climate change and environmental attitudes and behaviors: Insights and opportunities. *Frontiers in Psychology, 7*, 121.

Ferguson, M. A. *et al.* (2016). Global climate change: A social identity perspective on informational and structural interventions. In Mckeown, S., Haji, R., and Ferguson, N. eds., *Understanding Peace and Conflict Through Social Identity Theory: Contemporary and World-Wide Perspectives*. New York, NY: Springer.

Huttunen, J. and Albrecht, E. (2021). The framing of environmental citizenship and youth participation in the Fridays for Future movement in Finland. *Fennia, 199*(1), 46–60.

Johari, A. and Lalwani, V. (2021). Young climate activists in India are shaken — But proud they have rattled the powerful, *Scroll.in*, February 18, Climate Justice.

Kamal, N. (2022). Fridays for Future to hold international strikes Thursday calling for an end to war in Ukraine. *Times of India*, March 3. https://timesofindia.indiatimes.com/india/fridays-for-future-to-hold-international-strikes-thursday-calling-for-an-end-to-war-in-ukraine/articleshow/89966203.cms.

Kyroglou, G. (2021). An "inconvenient truth"? The problem of recognition of the political message — Commentary to Huttunen and Albrecht. *Fennia, 199*(1), 139–143.

Li, Y. *et al.* (2021). "Beyond clicktivism: What makes digitally native activism effective? An exploration of the sleeping giants movement. *Social Media + Society, 7*(3), 1–21.

Lohia A. (2020). Problems with draft EIA 2020: "Violator-friendly says Fridays for Future". *The Citizen*, August 6. https://www.thecitizen.in/index.php/en/newsdetail/index/13/19162/problems-with-draft-eia-2020-violator-friendly-says-fridays-for-future-india-.

Luomi, M. (2020). Global climate change governance: The search for effectiveness and universality, *International Institute for Sustainable Government (IISG)*, Brief#6.

Menon, V. ed. (2012). *Environment and Tribes in India: Resource Conflicts and Adaptations*. New Delhi: Concept Publishing Company.

Rees, J. H. and Bamberg, S. (2014). Climate protection needs societal change: Determinants of intention to participate in collective climate action. *European Journal of Social Psychology, 44*, 466–473.

Sherekar, S. (2022, February 15). Greta Thunberg's Fridays for Future attacks Pune's ambitious Mula-Mutha Riverfront Project: Here is all you need to know, *OpIndia:* https://www.opindia.com/2022/02/greta-thunbergs-fridays-for-future-attacks-punes-ambitious-mula-mutha-riverfront-project/

Swarnakar, P. (2019). Climate change, civil society and social movement in India. In Navroz. K. D. ed., *India in a Warming World*. Oxford: Oxford University Press.

Special Correspondent, Fridays for Future India comes out in support of Disha Ravi. *The Hindu*, February 20, 2021. https://www.thehindu.com/news/national/karnataka/fridays-for-future-india-comes-out-in-support-of-disha-ravi/article33884871.ece.

Vihma, A. (2011). India and the global climate governance: Between principles and pragmatism. *The Journal of Environment & Development, 20*(1), 69–94.

Von Zabern, L. and Tulloch, C. D. (2021). Changing thoughts, changing future — Commentary to Huttunen and Albrecht. *Fennia, 199*(1), 147–152.

Websites

www.fridaysforfuture.org
www.fridaysforfutureindia.com
Twitter:
Jairam Ramesh, @Jairam_Ramesh. February 14, 2021. Completely atrocious! This is unwarranted harassment and intimidation. I express my full solidarity with Disha Ravi. https://twitter.com/Jairam_Ramesh/status/1360903091332251651.

Greta Thunberg, @GretaThunberg. February 19, 2021. Freedom of speech and the right to peaceful protest and assembly are non-negotiable human rights. These must be a fundamental part of any democracy. #StandWithDishaRavi. https://twitter.com/GretaThunberg/status/1362776897436979208.

Chapter 13

Emerging Indian Partnerships in Climate Change with Special Reference to COVID-19 Era

Aditi Basu*

Abstract

The cyclones Amphan and Nisarga that hit India and Bangladesh in May 2021 have warned humanity of the dire consequences of climate change. Although many countries have taken a lead in deliberating and discussing on various issues of climate change, they have failed to take clear-cut decisions and joint action in reality. India has always viewed climate change as a diplomatic issue, justifying it by saying that the developed countries should probe into this matter because they are the ones who are responsible for causing the problem. Therefore, it is the responsibility of the First World Countries to call for global preponderance to address climate change.

Although the Global Change Data Lab has declared in 2021 that India is not responsible for the rising temperatures and sea levels, India believes that it needs to carve out a middle path on climate policy that could be accepted both in the domestic and international spheres. It is in the COVID-19 era that India has developed its climate action into climate diplomacy. The extreme climatic conditions like forest fires in

---

*An Independent Researcher.

U.S. and Australia, super cyclones in the U.S. and India, locust attacks in South Asia, and extreme heatwaves in the U.S., Europe, and Russia have wreaked havoc and led to the destruction of innumerable lives and property.

This chapter seeks to outline India's climate agreements with various countries in the Modi era with special reference to the Paris Agreement. It will also focus on India's stand and its increased awareness on climate change in the COVID-19 era. Special emphasis shall also be laid on India's climate agreements with the two power blocs, U.S. and China during this era as well as security issues concerning climate change.

Keywords: COVID-19, International Solar Alliance, Paris Agreement, United Nations, Security Issue, Sustainable Development

Introduction

Climate change is among one of the biggest challenges that nations across the world are confronted with. It is a compound of various economic and political interests of countries because the reduction efforts are solely dependent on the economic interests of countries. The occurrence of Hurricane Dorian, the forest fires of the Amazon and in Australia and the rampant destruction caused by them have emphasized the need for leadership and refocused on the fragility of the world's carbon sinks. Each country's response to climate change could "intrinsically contribute to international collaboration" (Halden 2007). The United Nations, playing an important role in combating climate change in the form of Intergovernmental Panel on Climate Change (IPCC), has raised international concerns on global warming. A multipolar world is, indeed successful in balancing for cooperation; however, it could also lead to serious divisions and conflicts without any arbiter to resolve them. Although the bulk of carbon emissions causing climate change are released by industrial countries, but their impacts are immensely felt in the poorer sections of the world (Giddens 2009). The political aspect of climate change implies the collective efforts of nations aspiring to "secure maximum emission liberty to ensure their developmental and economic goals aiming towards growth". The developing countries have been striving toward saving the basic differentiating structure of the United Nations Framework Convention for Climate Change (UNFCCC), but the Paris Agreement has given it a new meaning depending on differential national situations. Along with other rapidly industrializing countries, India has benefited from several years of high economic growth through a period of global economic slowdown, and is more assertively claiming a seat at the global high table. In the climate talks, this assertion has been facilitated by formation of the "BASIC" bloc of countries (Brazil, South Africa, India, and China). Significantly, in terms of per capita indicators of economic progress or greenhouse gas emissions, India has more in common with least developed countries than with the emerging rapidly industrializing economies, but through its own negotiation strategies and external perception tends to be increasingly identified with the latter rather than with the former.

India has become a key player of climate politics, becoming the third largest greenhouse gas (GHG) emitter. India aims to secure maximum emission space by not taking any legally binding emission reduction. The National Action Plan for Climate Change was established by India in 2009 and eight missions were set up to promote sustainable development and action on climate mitigation and adaptation, the center being "co-benefits".[1] An ex-member of the Prime Minister's Council on Climate Change (PMCCC) and India's lead negotiator at several climate summits, Chandrasekhar Dasgupta, labeled climate change as "threat multiplier" requiring "substantial global cooperation not only for mitigation of climate change but also to adapt to the impact of climate change". India has developed a protest voice on global climate policy which actively raises international concerns for global environmental efforts (Michaelowa and Michaelowa 2012). India's role in the success of Paris Agreement has been praised by many countries.

Climate Change as an Issue of Indian Foreign Policy in the Modi Era

India's foreign policy has been advocating for a climate-resilient development agenda through its bilateral, regional, and multilateral collaborations which can enhance trade prospects, investments to inclusive solutions, thereby, contributing to sustainable development.

India is one of the signatories to the Paris Agreement under the United Nations Framework Convention on Climate Change (UNFCCC) which was adopted at the UN climate conference "CoP 21" (Conference of

---

[1]Climate Change Programme. *Ministry of Science and Technology*, February 25, 2021. https://dst.gov.in/climate-change-programme.

Parties) held in 2015, the motive being collective action toward reduction of hazardous greenhouse gas emissions. The Paris Agreement, an international treaty adopted by 196 parties at the 21st Conference of Parties in Paris, on December 12, 2015 is a legally binding agreement for the countries to take collective action on climate change. The Agreement came into force on November 4, 2016 (Soni 2020). Ever since the adoption of the Paris Agreement, there have been debates on the initial insufficient commitments made by more than 185 countries to fight climate change. The target year for the fulfillment of the first round of more influential climate plans was set as 2020 (Kerry 2019).

India led the International Solar Alliance (ISA) in the Paris Climate Change Summit in 2015 for the promotion of solar energy which is a clean and renewable energy to humanity. The ISA, a treaty-based initiative by both India and France, may be useful in playing a key role in the enhancement of low-carbon transitions among its member countries: 90 countries have signed ISA's framework agreement and 73 countries have ratified it. This could be a significant achievement for the developing countries if Indian Prime Minister Narendra Modi is successful in "obtaining the adherence of all the participating countries in this ambitious project, the U.S. and the European Union as a community in particular" (Saran 2021) and could also set the Indian foreign policy in the right direction. The ISA could strengthen India's diplomatic ties with the continental engagement Initiatives like the African Union and Indian Ocean Rim Association (IORA) and with other member countries of the ISA. The period between 2015 and 2019 was critical due to the-then U.S. President Donald Trump's disavowal of the Paris Agreement and this significant change led to China and India trying to develop themselves as responsible powers, "hinging on multilateralism and the rule of law" (Jayaram 2021).

India has become the world leader in the solar and wind sectors, ranking fifth and fourth, respectively, in cumulative capacity installations in 2019,[2] and has gained largely in the improvement of energy efficiency in its economy. India has also assured a sum of $26 million for creating a corpus fund for ISA and has "opened lines of credit worth $1.39 billion for implementing 27 projects in 15 ISA member countries, 13 of which are African states" (Shaji and Susarla 2021). India and the Association of Southeast Asian Nations (ASEAN) have agreed jointly for reducing the

---

[2]REN21 (2020). Renewables 2020 — Global Status Report. https://www.ren21.net/wp-content/uploads/2019/05/gsr_2020_full_report_en.pdf.

adversities of climate change, increasing livelihood opportunities for the people by the promotion of agro-forestry, exchanging farm machineries, and developing heterotic rice hybrids. The joint declaration, signed by both parties, said, "We look forward to cooperation on exchanging expertise as to promote enhanced resilience on natural systems and improve the adaptive capacities of people to cope with environmental hazards".[3]

India along with the South Asian Association for Regional Cooperation (SAARC) countries (Bangladesh, Sri Lanka, Pakistan, Afghanistan, Nepal, Maldives, and Bhutan) has pledged for the adoption of technologies that would reduce the impact of climate crisis on fisheries and aquaculture and has also sought for regional cooperation for cross-learning for the sustainable utilization of the resources.[4] In the brochure launch of India's climate policy on June 5, 2020, India's External Affairs Minister S. Jaishankar had pointed out that India's stand in addressing climate change is firm. Apart from highlighting the "consequences of melting of glaciers on the deployment of troops at high-altitude locations on India's mountainous borders", Jaishankar also focused on "large scale migration of population from low-lying coastal plains towards higher ground" leading to "social disruptions and economic distress, undermining domestic security" (Saran 2021). In the G-20 Summit, PM Modi said, "India is not only meeting targets set at the Paris Agreement but is also exceeding them while calling for 'an integrated, comprehensive and holistic way' to tackling climate change".

During the launch of Green Sohra plantation drive by Assam Rifles in Meghalaya, Hon'ble Indian Minister of Home Affairs, Amit Shah said, "Global warming and climate change has become a crisis and in the fight against it, PM Modi is taking a lead role by launching several schemes by distributing free gas supply to decrease the carbon footprint" (Choudhury 2021). On August 27, 2021, PM Modi launched 35 crop varieties with "special traits like climate resilience and higher nutrient content" and urged academicians, agricultural scientists, and institutions

---

[3] ASEAN–India vow to tackle climate change and price volatility. *The Livemint*, June 12, 2018. https://www.google.com/amp/s/www.livemint.com/Politics/oJmTa6s68M6Ttk5dxX-1CgP/AseanIndia-vow-to-tackle-climate-change-and-price-volatilit.html%3ffacet=amp.
[4] SAARC nations for technology use to fight climate crisis in fisheries. *The Business Standard*, August 6, 2021. https://www.google.com/amp/s/wap.business-standard.com/article-amp/pti-stories/saarc-nations-pitch-for-use-of-tech-in-reducing-climate-crisis-in-fisheries-acquaculture-121080501428_1.html.

to create awareness on climate change among farmers through campaigns (Sharma 2021).

Evolution of India's Climate Diplomacy

A close and careful analysis of India's climate diplomacy shows that India is committed toward securing equity, climate justice, reducing vulnerability, poverty eradication, and development. Since 2010, there have been significant transitions and shifts in India's climate diplomacy and multilateral, bilateral, and informal exchanges on climate change cooperation have increased. Since 2014 when the Modi-led Bharatiya Janata Party (BJP) came to power, the Indian foreign policy has shown remarkable trends. It has been more decisive and has been renewed with a spirit of new vigor and energy, thanks to Indian Prime Minister Narendra Modi (Sidhu and Godbole 2015). This has created sufficient grounds for India to gain leadership towards the Paris Agreement in Indian climate policy (Dubash 2009). This paradigm shift has occupied the center of India's overall diplomatic strategy and has been revitalized under the present BJP-led government. The present Modi government has included climate diplomacy in its foreign policy agenda too. The integration of the ISA and the Coalition for Disaster Resilient Infrastructure (CDRI) into the multilateral climate action agenda is a clear example of the paradigm shift (Jayaram 2021).

Although the decision to adopt voluntary targets at the Copenhagen Summit — such as to reduce the emissions intensity of its Gross Domestic Product (GDP) by 20–25% against 2005 levels by 2020 — could be regarded as one of the first signs of shift in India's positions in the international climate change negotiations, it is only in the run up to the 2015 Paris Summit (21st Conference of Parties or CoP 21) that India's willingness to be a global climate leader became apparent. India has developed its stand on climate diplomacy since 1990s by highlighting the issue of environmental colonialism through the principle of "Common but Differentiated Responsibilities" (CBDR) that has led to the establishment of the ISA in 2015. Since India's climate goals are connected with its energy commitments, its climate diplomacy deals with tackling both energy security and climate change.

India has stressed on the need for "common but differentiated responsibilities in international cooperation" for addressing climate change.

This includes collective climate action on the part of the developed countries to ramp progress which could substantially increase new climate finance and enable more technology transfers to developing countries as features of India's climate diplomacy.[5] India has also laid stress on the fact that "global development, addressing climate change, and eradicating poverty are central to the planet's future".[6] Thus, the ideas of climate justice and inclusive transitions form the foundation stone of the country's climate diplomacy. India could utilize its opportunity of setting and leading a development-centered, climate, and clean energy agenda. This could be achieved by increasing its domestic ambitions and international engagements in the sphere of sustainable policies of climate change and also by the boosting multilateral, regional, and bilateral diplomatic efforts, thereby, working toward inclusive low-carbon transitions and developed climate resilient economies and communities to climate finance. India aims to carve out its own ideas of climate action and justice by laying stress on non-traditional security challenges and by playing a positive role in global politics.

In September 2019, India announced that it would proceed with the project of launching massive "175 Gigawatts of new installed renewable capacity by 2022" (Kerry 2019). India has also argued that "any commitment to net-zero would mean compromising developmental goals of countries with a far shorter legacy of emissions compared to the developed world". New Delhi has highlighted the poor track record of the developed countries in fulfilling their commitments of technology transfer and financial aid to developing countries.[7] In 2019, the Dominican Republic had convened an open debate on the effects of climate-related disasters on international peace and security. In this, the Indian representative to the UN Syed Akbaruddin had said that the "security approach" to climate change could hinder "the global collective effort" to tackle

---

[5] Ministry of Finance (2019). *Climate Summit for Enhanced Action: A Financial Perspective from India*. New Delhi: Government of India. https://dea.gov.in/sites/default/files/FINAL%2017%20SEPT%20VERSION%20Climate%20Summit%20for%20Enahnce%20Action%20A4%20size.pdf.

[6] External Affairs Minister's remarks during launch of India's priorities for its candidature for the UN Security Council 2021–22.

[7] IPCC report could intensify pressure on Delhi to hike its climate ambition. *The Indian Express*, August 11, 2021. https://www.google.com/amp/s/indianexpress.com/article/opinion/editorials/ipcc-climate-change-report-7447887/lite/.

climate challenges and that it was not the correct governance mechanism to tackle a "global challenge" like climate change. In 2019, India, China, and France have jointly decided to frame their climate policies "in a manner representing a progression beyond the current one and reflecting their highest possible ambition" and to "cut down on emissions in the long run" as enumerated under the Paris Agreement (Kerry 2019). India also introduced clean energy goals in its domestic policies to tackle India's increasing climate risks in the short and medium terms.

> "Under the Paris Agreement, India has three quantifiable nationally determined contributions (NDCs), which include lowering the emissions intensity of its GDP by 33-35 per cent compared to 2005 levels by 2030; increase total cumulative electricity generation from fossil free energy sources to 40 per cent by 2030 and create additional carbon sink of 2.5 to 3 billion tons through additional forest and tree cover".[8]

On September 23, 2021, in the United Nations Security Council (UNSC) debate on "Climate & Security", India highlighted the bottlenecks of focusing only on one aspect (security) of climate change while ignoring the others and advocated for a more comprehensive and collective approach. India was represented by Secretary (West) in the Ministry of External Affairs Reenat Sandhu who stressed on the urgency of immediate action with respect to climate change based on the principle of "Common but Differentiated Responsibility" and Respective Capabilities. She also said that highlighting climate security in the platform of the Security Council and "ignoring the basic principles and practices relating to climate change" was not a good option at all. India was of the opinion that viewing conflicts in the underdeveloped regions of the world through the lens of climate change would provide only a "lop-sided narrative" although some other regions were responsible for the conflicts. Sandhu held that climate change could enhance conflicts and hence, the parties need to be conscious while viewing climate change as an issue which could risk their social stability, peace, and security. India has been working toward combating climate change through expansion of solar energy

---

[8] India's intended nationally determined contribution. https://vikaspedia.in/energy/environment/climate-change/indias-intended-nationally-determined-contribution#:~:text=To%20adopt%20a%20climate%2Dfriendly,by%202030%20from%202005%20level.

program and clean cooking fuel to cover over 80 million households. Added to it, 370 million LED light bulbs have been distributed which has resulted in an energy saving of more than 47 billion units of electricity per year and a reduction of more than 38 million tons of CO_2 emissions annually. She held that India is committed to install 450 GW of new and renewable energy by 2030. Apart from the domestic initiatives, India has also decided to lead all forms of international partnerships for generating long-term impact through coalitions in the form of ISA and CDRI (Singh 2021).

India's Multilateral Climate Agreements in the COVID-19 Crisis

Due to the COVID-19 pandemic, the country's economy is facing various uncertainties. India's approach to multilateral action in its foreign policy has five pillars namely the five S's — *Samman* (respect), *Samvad* (dialogue), *Sahyog* (cooperation), *Shanti* (peace) *and Samriddhi* (prosperity)[9] as articulated by India's External Affairs Minister, Jaishankar in his speech at the launch of India's UNSC Campaign Brochure. This has revealed India's multilateral approach on climate change committed to rule of law and a "fair and equitable international system". The government has also recognized the emergence of "new and complicated challenges" requiring a "coherent, pragmatic, nimble and effective platform" for mutual collaboration ensuring sustainable peace". India's recent experience of dealing with COVID-19, the cyclones Amphan and Nisarga, a locust attack and extreme heat, among others, have, in a way, "underscored the interconnected nature of several risks that have implications for governance and security".

While delivering the 19th Darbari Seth Memorial Lecture in 2020, the UN Chief bestowed on India the responsibility of leading the world in spheres of energy and health and to create "inclusive economies and avert the threat of climate change". He also said that if India is successful in doing so, it could truly become a global superpower[10] in fighting climate

---

[9] India's External Affairs Minister S. Jaishankar quoting Prime Minister Modi in S. Jaishankar, "Remarks During Launch of India's Priorities for its Candidature for the UN Security Council 2021–22" (Speech, Ministry of External Affairs, June 5, 2020).

[10] *UN Secretary General Delivers TERI's 19th Darbari Seth Memorial Lecture*. Mithapur: TERI. https://www.teriin.org/un-secretary-general-delivers-teri-19th-darbari-seth-memorial-lecture.

change. For this, India needs to speedily shift from fossil fuels to renewable energy. In the G-20 summit titled "Safeguarding the Planet: The Circular Carbon Economy Approach", Modi said that India is not only meeting its Paris Agreement targets, but also exceeding them.[11] India occupies a "grey zone between the developing and developed worlds" with occupying an important stand in G77 on one hand and another partial stand in G7. India shares warm strategic relationships with most G7 countries while sticking itself to its position of being one of the third world countries. While it actively works toward the attainment of climate goals for itself, along with that, it also urges the developed countries to enhance collective action for adaptation and mitigation (Pant 2021). Under the Copenhagen Accord of 2009, the Indian government had pledged for the reduction of emission intensity levels by 20–25% since 2005 till 2020 (Deshmukh 2010). The developed countries also had pledged to mobilize $100 billion a year by 2020 to help the developing countries in mitigating the adversities of climate change.[12]

When Joe Biden became the President of the United States, he appointed John Kerry as the climate envoy, which has created a visible global shift in climate action discourses. The recent revival of QUAD has revitalized climate change as an "indispensable action" involving Australia, India, Japan, and the U.S. (Jayaram 2021). Hon'ble U.S. President Joe Biden had convened a virtual summit on climate change with 20 top world leaders on April 2021 which contribute to 80% of the global carbon emissions. In April 2021 at the virtual Leaders' Summit on Climate Change convened by the U.S. President Joe Biden, Prime Minister Narendra Modi announced the Indian collaboration with the U.S. for the mobilization of investments, promotion of cooperation to combat the global warming, demonstrating clean technology and enhance green collaboration. He also reminded that the COVID-19 is the result of climate change. Both U.S. and India launched the India–U.S. Climate and Clean Energy Agenda 2030 Partnership. Meanwhile, India made it clear that it is willing to engage in partnerships for creating a sustainably

---

[11] G20 Summit. PM Modi calls for integrated approach to combat climate change. *The Hindu*, November 22, 2020. https://www.google.com/amp/s/www.thehindu.com/news/national/climate-change-must-be-fought-not-in-silos-but-in-integrated-holistic-way-pm-modi-at-g20/article33155391.ece/amp/.

[12] *Climate Finance in the Negotiations*. New York: UNFCCC. https://unfccc.int/topics/climate-finance/the-big-picture/climate-finance-in-the-negotiations.

developed India which could also help other developing countries to access green finance and clean technology. India has also launched global initiatives like the ISA and the CDRI in which 18 countries (9 among the G20 and four international organizations have already joined) (Louis 2021). This has promoted a spirit of cooperation and collaboration among all the developing countries. The U.S. is a founding member of the Indian initiative of CDRI and will work for raising "private and public sector financing for infrastructure that can stand upto natural disasters". United States Agency for International Development (USAID) has also started a five-year initiative of mobilizing $7 billion in private clean energy in South Asia, most of which covers India. USAID is also partnering with the U.S. Department of Energy to launch the South Asia Group for Energy to support India.[13]

At the Earth Day climate summit, India and China jointly held a virtual meet of 40 world leaders in which Biden pledged to "pledged to cut the US's greenhouse gas emissions by 50% by 2030, doubling the US's goals under the Paris Climate Accord". However, both Chinese President Xi Jinping and Indian Prime Minister Narendra Modi refrained from making any decisions on climate crisis. Xi Jinping announced to "strictly control" coal-fired plants and to "limit the increase in coal consumption until 2025" under China's Five-year plan. He also proposed to focus on building "cooperation on ecological civilization in the joint building of the Belt and Road Initiative" which would prove to be beneficial to all participating countries, connecting 70 countries across Asia, Africa, and Europe through a web of land and sea networks. This ecological civilization would work toward sustainable development, low-carbon framework. Xi said, "We must join hands, not point fingers at each other; we must maintain continuity, not reverse course easily; and we must honour commitments, not go back on promises". He also said that although China was facing "immense difficulties" in restructuring its economy, even as it was treading faster toward carbon neutrality than U.S.A. and Europe. He said that Chinese and American joint working groups would continue to agree on climate change. In this summit, Modi laid stress on the importance of "sustainable development" (Teh 2021).

---

[13] India critical part of global climate change solutions, says USAID Administrator. *The Indian Express*, October 6, 2021. https://www.google.com/amp/s/www.newindianexpress.com/world/2021/oct/06/india-critical-part-of-global-climate-change-solutions-says-usaid-administrator-2368459.amp.

In the Quad virtual meeting of India, Australia, Japan, and the U.S. in March 2021, areas such as the Quad vaccine partnership, critical and emerging technologies and climate change were discussed. The leaders of these countries decided to work jointly on the distribution of COVID-19 vaccines and climate change which was an effort to counter the growing Chinese hegemony (Laskar 2021).[14] Through the Quad, Indian External Affairs Minister S. Jaishankar and U.S. Secretary of State Antony Blinken discussed the supply of COVID-19 vaccines to India, strengthening the Indo-Pacific cooperation through the Quad alliance and fighting climate crisis by enhancing multilateral cooperation.[14]

Climate change has been an important strategic area for fresh collaboration between India and Bangladesh too since the last two decades (Khan 2021). These two countries should "unite to jointly address such ferocious climate impacts and to build their adaptive capacity and resilience". India has realized the geopolitical implications of climate change with special reference to being one of the most important countries of South Asia and the need for the adoption of a "climate security lens addressing climate-related challenges". This could be achieved through broader regional cooperation and greater flexibility on climate security. India has proved itself in the international arena that it is well positioned in promoting dialogue with the developing countries on many issues, security implications of climate change being one of them.[15]

India's security and climate interests could be affected by China's growing strategic interests in the Arctic for gas, oil, rare earths, and ship routes and its interest in creating an alternative source of petroleum in the Arctic. India and China have been "observer" states in the Arctic Council since 2013. The Director of the National Institute of Advanced Studies, Shailesh Nayak has said, "China's main interest is strategic and economic development of the Arctic as well as climate change, and has clear policy on these aspects. China is keen to develop the northern Arctic Sea Passage as a shipping route and related infrastructure for economic considerations,

---

[14] Jaishankar, Blinken discuss vaccine supplies, climate change, Quad Alliance. *The Hindu Business Line*, May 29, 2021. https://www.google.com/amp/s/www.thehindubusinessline.com/news/national/jaishankar-us-defence-secy-discuss-shared-priorities-regional-security-challenges/article34672941.ece/amp/.

[15] *Will India Shift Its Stance on Climate Security in the UNSC?* Den Haag: Planetary Security Initiative. https://www.planetarysecurityinitiative.org/news/will-india-shift-its-stance-climate-security-unsc.

as evidenced by its participation in the Arctic Economic Council and Arctic Circle forum. The route also gives a military advantage to China, compared to politically unstable regions" (Chauhan 2021).

On September 23, 2021, while interacting with top business leaders in the U.S., Indian PM Modi met First Solar CEO Mark Widmar to discuss Indian policies for climate change and related industries, according to the sources. PM Modi spoke about "India's ambitious target of 450 GW of renewable energy and expanding solar energy" which could be useful for the companies in the related field under the PLI (Production Linked Incentive) schemes. Both of them agreed for enhancing the manufacture of solar energy in India benefiting the neighboring countries. After the interaction, Widmar said, "With his leadership and what he has done to create a really strong balance between industrial policy as well as trade policy, makes it an ideal opportunity for companies like Solar, to establish manufacturing in India, and his commitment to ensuring domestic capabilities and ensuring his long term climate goals and objectives with focus on energy independence and security. It is not one that we would forego, we would want to be part of this opportunity, to be part of this journey". He also commented that the Indian initiative of ISA would help U.S.A. to "think about opportunities" and U.S.A. would want to be a technology leader in India's levering capabilities.[16] India and Australia have decided to hold the inaugural "2+2" Ministerial meeting during the September 2021 visit of Foreign Minister Marise Payne and Minister of Defense Peter Dutton. Important issues as economic, cyber security, climate change, critical technology, and supply chains were discussed (Bhattacharjee 2021).

In this summit, India has welcomed the efforts undertaken by the U.S.A. in combating climate change and in rejoining the Paris Agreement when Indian Prime Minister Narendra Modi and U.S. President Joe Biden held a bilateral meeting at the White House. The U.S. decided to support India in its drive to achieve its domestic goal of installing 450 GW of renewable power by 2030. Ahead of the 26th UN Climate Change Conference of the Parties (CoP 26) in U.K.'s Glasgow, this meeting was significant. During this meeting, Biden focused on "mobilising finance

---

[16]All countries should emulate what India has done in climate change: First solar CEO. *The Times of India*, September 23, 2021. https://www.google.com/amp/s/m.timesofindia. com/india/all-countries-should-emulate-what-india-has-done-in-climate-change-first-solar-ceo/amp_articleshow/86461226.cms.

for investments in renewables, storage and grid infrastructure" guaranteeing clean and non-conventional sources of power equipping millions of households. India and the U.S. have partnered each other to develop clean energy and deploy critical technologies through the two most important tracks: Strategic Clean Energy Partnership (SCEP) and the Climate Action and Finance Mobilization Dialogue (CAFMD).

India has reiterated that it is encountering climate- related issues as a result of follies that have been committed by developed nations, responsible for the climate change. Hence, these countries should help the developing world by making technologies an affordable cost. They have committed to achieve a goal of mobilizing $100 billion a year with 2020 being the deadline. In August 2021, Hon'ble Environmental Minister Bhupender Yadav said, "India is committed to the UNFCC and its Paris Agreement and had extended support to the UK, which will host the international climate conference".[17]

The 26th Conference of Parties (CoP 26) was held from October 31 to November 12 in 2021 in Glasgow, U.K. This is one of the most important summits of climate diplomacy since 2015, when the Paris Agreement failed to meet the requirements of 197 members of the UNFCCC. According to the IPCC, all the countries have set their target of 1.5°C of warming by 2040, majority being the South Asian countries. They have been pressurizing the Modi administration to set a deadline of 2050, a time by which India would reach "net-zero" emissions, which means that it would be able to absorb all the emissions it produces. India did decide to release its new climate policies at CoP 26 (Bello 2021).

On October 6, 2021, USAID Administrator Samantha Power said that India is "a critical part of the global climate change solutions". According to her, India is not only "at the mercy of a changing climate" but also a part of the solution. Climate change has impeded the growth of the Indian economy during the COVID-19 pandemic. A rise of 1°C in the global temperature could lead to a fall in 3% of Indian GDP.[18]

---

[17]*Climate Finance in the Negotiations*. New York: UNFCCC. https://unfccc.int/topics/climate-finance/the-big-picture/climate-finance-in-the-negotiations.
[18]India critical part of global climate change solutions, says USAID Administrator. *The Indian Express*, October 6, 2021. https://www.google.com/amp/s/www.newindianexpress.com/world/2021/oct/06/india-critical-part-of-global-climate-change-solutions-says-usaid-administrator-2368459.amp.

On October 9, 2021, at the UN General's General Debate of the Second Committee on "Crisis, Resilience and Recovery — Accelerating Progress towards the 2030 Agenda", India's Permanent Representative to the UN Ambassador T.S. Tirumurti said that the developed countries should do a "Net-Minus to vacate space for the developing countries to grow" by 2050 and by 2030 they should reduce their emissions to half based on the principle of equity which means that no new emissions should be added to the atmosphere by the developed countries so that developing countries like India can grow economically and industrially.[19]

Conclusion

India enjoys the status of being a signatory to the Paris Agreement that was ratified on October 2, 2016 and enforced on November 4, 2016. India has gained the third position on the highest emission of greenhouse gases (4.10%), China and the U.S.A. being in the first and second positions. India is paying more attention to the climate change policy as there has been a 1°C rise in global temperatures which is quite a serious issue, leading to frequent droughts, hurricanes, forest fires, cyclones, and other natural disasters. Even a rise in 0.5°C in global temperature could severely impact developing countries. India views climate change as an issue of immediate concern and if the temperature rise is kept in limit to only 1.5°C instead of 2°, this could save the lives of millions of people, especially in a developing country like India. The risks would impact poverty due to limited agricultural production because of temperature rise leading to food scarcity. Although India has clearly mentioned that it is suffering the adversities of climate change due to the rapid industrialization of the developed countries, it has called for global preponderance of climate finance and technology at an affordable cost.

The U.S. and China, although rivals, could collaborate with each other as far as climate change and energy security are concerned because they "hold the future in hands" (Giddens 2009). According to Shyam Saran,

---

[19]Developed countries should do a Net-Minus to vacate carbon space in 2050 for developing countries to grow: India. *The Indian Express.* https://www.google.com/amp/s/indianexpress.com/article/india/developed-countries-net-minus-emissions-carbon-space-2050-paris-agreement-climate-change-un-7562018/lite/.

India should guarantee "green recovery" by laying stress on alternative economic strategies of sustainable growth. This would, in turn, place India in a position to achieve the status of low-emission growth as compared to the announcement of ambitious targets for the achievement of what is missing in India's current climate policy. Saran also called for the integration of the current situation of climate change with future security frameworks such as military, social, and political security (Saran 2021). India's position in the formulation of a world opinion and an institutional build-up on climate change shows a broader developmental approach focused on adaptive capacity and risk assessment. India's climate diplomacy needs to focus more on development and its approach should be multifaceted. This multifaceted approach should focus on adaptation and encourage more Western engagement on issues like finance and technology to tackle climate change. India needs to frame its national priorities more precisely so that it earns more worldwide cooperation. Greater global cooperation in technology and advanced scientific research could help in tackling the climate crisis.

If India is able to engage in more multilateral cooperation with climate leadership and focus on sustainable and planned addressing of its own climate vulnerabilities, it could soon rise above the status of a developing country. Modi has rightly spoken about "behavioral change" as the best way to "fight climate change" (Jayaram 2021). India's climate diplomacy is no longer confined to the Copenhagen Accord and Paris Agreement. India has undergone a considerable transition in its climate diplomacy. The dormant Indian Ocean Rim Association (IORA) could be revived as it has the potential to protect the region's climate interests (Gargeyas 2021). Increased multilateralism, promoting global cooperation on climate change is the need of the hour. The role of climate politics in India's foreign policy has created a paradigm shift in Indian climate policy. This shift in India's overall diplomatic strategy had initially begun during the financial crisis in 2008 but it has been re-energized and revitalized under the Modi-led Indian government.

References

Bello, L. D. (2021). What's at Stake for India and South Asia at COP 26? *The Wire*, October 1. https://science.thewire.in/environment/whats-at-stake-for-india-and-south-asia-at-cop-26/.

Bhattacharjee, K. (2021). India, Australia to hold 2+2 meet. *The Hindu*, September 8. https://www.google.com/amp/s/www.thehindu.com/news/national/india-australia-to-hold-22-meet/article36369867.ece/amp/

Chauhan, B. (2021). China's growing Arctic footprint may hit India's security and climate interests: Experts. *The New Indian Express*, August 13, 2021. https://www.google.com/amp/s/www.newindianexpress.com/nation/2021/aug/13/chinas-growing-arctic-footprint-may-hit-indias-securityand-climate-interests-experts-2344025.amp.

Choudhury, R. (2021). PM Modi in lead role in fight against climate change: Amit Shah. *NDTV*, July 25. https://www.google.com/amp/s/www.ndtv.com/india-news/pm-modi-in-lead-role-in-fight-against-climate-change-amit-shah-2494443%3famp=1&akamai-rum=off.

Deshmukh, R., Gambhir, A., and Sant, G. (2010). Need to realign India's national solar mission. *Economic and Political Weekly, 45*, 10.

Dubash, N. K. (2009). *Toward a Progressive Indian and Global Climate Politics*. New Delhi: CTCN.

Gargeyas, A. (2021). Climate change is the biggest threat to Indian Ocean security. *The Diplomat*, August 31, 2021. https://thediplomat.com/2021/08/climate-change-is-the-biggest-threat-to-indian-ocean-security/.

Giddens, A. (2009). *The Politics of Climate Change*. London: Polity Press.

Halden, P. (2007). *The Geopolitics of Climate Change*. Stockholm: Swedish Defence Research Agency.

India just as India supported US earlier, says Blinken. *The Hindu Business Line*, May 29, 2021. https://www.google.com/amp/s/www.thehindubusinessline.com/news/national/jaishankar-us-defence-secy-discuss-shared-priorities-regional-security-challenges/article34672941.ece/amp/.

India welcomes US' efforts to fight climate change, its return to Paris Agreement. *The Kashmir Reader*, September 26, 2021. https://www.google.com/amp/s/kashmirreader.com/2021/09/26/india-welcomes-us-efforts-to-fight-climate-change-its-return-to-paris-agreement/amp/.

Jayaram, D. (2021). India's climate diplomacy has to move up a gear. *India Climate Dialogue*, March 19. https://indiaclimatedialogue.net/2021/03/19/indias-climate-diplomacy-needs-to-move-up-a-gear/.

Kerry, J. (2019). *China and India Must Step up on Climate Change*. Washington, D.C.: Carnegie Endowment for International Peace. https://carnegieendowment.org/2019/09/22/china-and-india-must-step-up-on-climate-change-pub-79897.

Khan, M. R. (2021). The implications of climate change for Bangladesh–India relations. *The ORF*, March 25. https://www.google.com/amp/s/www.orfonline.org/expert-speak/implications-climate-change-bangladesh-india-relations/.

Laskar, R. (2021). Vaccine partnerships, climate change on Quad, key officials meet. *Hindustan Times*, August 12. https://www.google.com/amp/s/www.

hindustantimes.com/world-news/vaccine-partnerships-climate-change-on-agenda-of-quad-key-officials-meet-101628785847521-amp.html.

Louis, A. (2021). Modi announces US-India partnership to fight climate change. *The Quint*, April 23. https://www.google.com/amp/s/www.thequint.com/amp/story/news/hot-news/modi-announces-us-india-partnership-to-fight-climate-change-2.

Michaelowa, K. and Michaelowa, A. (2012). India as an emerging power in international climate negotiations. *Climate Policy*, pp. 575–590.

Pant, H. V. (2021). The United States and India: Multilaterally abridged allies. *The ORF*, August 24. https://www.google.com/amp/s/www.orfonline.org/research/the-united-states-and-india-multilaterally-abridged-allies/.

Prunier, G. (2005). *Dafur, the Ambiguous Genocide*. London: Hurst.

Saran, S. (2021). India doesn't have to match climate commitments expected of China. Modi must make it clear. *The Print*, April 21. https://www.google.com/amp/s/theprint.in/opinion/india-doesnt-have-to-match-climate-commitments-expected-of-china-modi-must-make-it-clear/643132/%3famp.

Shaji, S. and Susarla, A. (2021). India, Africa, and climate diplomacy. *The Indian Express*, February 14. https://indianexpress.com/article/opinion/india-africa-and-climate-diplomacy-7188554/.

Sharma, H. (2021). Climate change big challenge for entire ecosystem, need to step up efforts: PM Modi. *The Indian Express*, September 28. https://www.google.com/amp/s/indianexpress.com/article/india/pm-narendra-modi-climate-change-millets-international-market-nep-7539244/lite/.

Sidhu, W. P. S. and Godbole, S. (2015). Bold initiatives stymied by systemic weakness. In Sidhu, W. P. S. and Godbole, S. eds., *Modi's Foreign Policy @365: Course Correction*. New Delhi: Brookings India.

Singh, S. (2021). India against confining Climate Change to Security, pushes for "comprehensive approach". *republicworld.com*, September 23. https://www.google.com/amp/s/www.republicworld.com/amp/india-news/general-news/india-against-confining-climate-change-to-security-pushes-for-comprehensive-approach.html.

Soni, M. (2020). 5 Years On, where India stands in its commitment to Paris Agreement. *The Hindustan Times*, December 12. https://www.google.com/amp/s/www.hindustantimes.com/india-news/5-years-on-where-india-stands-in-its-commitment-to-paris-agreement/story-M31plpICVbCP264U3q2h5L_amp.html.

Teh, C. (2021). China and India, 2 of the world's biggest greenhouse gas polluters, stay silent on new emissions targets at global climate summit. *Insider*, April 22. https://www.google.com/amp/s/www.insider.com/china-and-india-stay-silent-on-new-emissions-targets-at-earth-day-summit-2021-4%3famp.

Index

A

Adaptation Action Coalition, 17
aesthetics, 55
affordable finance at scale, 156
Agent Orange, 84
Alliance of Small Island States
  (AOSIS), 182
alternative energy grid, 164
Amazon, 236
Anthropocinema, 51
Anthropos, 28
anti-satellites weapons, 108
apple farmers, 60
Armageddon, 46
Assam Rifles, 239
augmentation, 123
AUKUS, 76
Azadi ka Amrit Mahotsav, 8

B

Bangladesh, 12
blue marble image, 29
Bolsonaro administration, 84
Boserupian, 26
Bundelkhand, 59
Buriganga river, 86

C

Carbon Border Adjustment
  Mechanism, 138
Carbon Offsetting and Reduction
  Scheme for International Aviation
  (CORSIA), 17, 190
CEEW, 8
China's Long March 5B, 118
Chinese Beidou Satellite Navigation
  Systems, 109
Chinese fishing vessels, 75
circular economy, 134
clean energy, 133
Climate Change Securitisation
  Paradox, 105
Club of Rome Report, 29
Coalition for Disaster Resilient
  Infrastructure (CDRI), 7
Cold War, 27
common-but-differentiated
  Responsibilities, 10
COP26, 2
Copenhagen, 15
coral bleaching, 55
coral reefs, 53
CSR funds, 98

CubeSats, 116
Curtisian aesthetic, 61
cyclones, 236

D
Dhaleshwari river, 86
documentary, 53
droughts, 60
dystopia, 54

E
eco-criticism, 51
ecological footprint, 122
Elite Theory, 87
Elon Musk's Space X, 116
emission reduction, 173
energy efficiency, 238
enlightenment, 55
environmental laws, 84
envirotoon, 51
epistemologies, 61
EU–Africa Summit, 145
Eurocentric, 33
European Climate, 133
European energy regulators, 144
European Green Deal, 14, 131
European Investment Bank, 156
eutrophication, 97
exclusive economic zones (EEZs), 11

F
Facebook, 222
fifth continent, 77
First World, 19
fluorescing, 56
Free Rider, 173
Friday for Future (FfF), 18

G
Game theory, 177
Gazprom, 141

general catalog of artificial space
    objects, 112
Global Change Data Lab, 19
Global Climate Change Index 2022,
    6
global gateway, 145
Global Methane Pledge, 8, 197
global south, 4
Gothic response, 61
Great Acceleration of the
    Anthropocene, 55
green agenda, 79
Green Growth and Equity Fund,
    212
greenhouse effect, 119
greenhouse gas (GHG) emissions, 4
Greta Thunberg, 47

H
Habermasian, 174
heterodystopias, 38
Himachal Pradesh, 60
Horizon Europe, 138

I
Indian chapter, 221
Indian Ocean, 77
Indian Ocean Rim Association, 238
India's Coastal Communities, 212
Indira Gandhi, 29
Indo-Burma Biodiversity Hotspot,
    85
Indo-Pacific, 71
Indo-Pacific Carbon Offsets, 76
institutions, 109
intended nationally determined
    contribution (INDC), 10
Inter-Agency Space Debris
    Coordination Committee, 110
Intergovernmental Panel on Climate
    Change, 4

International Monetary Fund (IMF), 31

International Solar Alliance (ISA), 7, 151

K

Kaptai Dam, 85
Kendrapara, 59
Kerala floods, 59
Kiribati, 75
Kyoto, 15
Kyoto Protocol, 31

L

landscapes, 58
LinkedIn, 222
Lithium batteries, 164
locust attacks, 236
Lower Earth Orbit (LEO), 13, 106

M

Madagascar, 57
Mahatma Gandhi, 228
Malthusian, 29
Manhattan, 57
Manus naval base, 79
Mars, 118
metaphors, 29
Mission Innovation, 194
Montreal Protocol 1987, 172
Moulvibazar, 85
mythologies, 49

N

NASA, 112
National Action Plan on Climate Change (NAPCC), 17, 201
National Water Policy, 205
neo-realists, 33
Non-Aligned Movement, 189
Nord Stream 2, 141

North–South divide, 3
Nuclear Suppliers' Group, 190
number problem, 182

O

Oceania countries, 70
Odisha, 59
Official Development Assistance (ODA), 11
One Sun, One World, One Grid (OSOWOG), 8

P

30by30 pledge, 17
Pacific Island Countries, 11, 67
Pacific Marine Industrial Zone (PMIZ), 75
Pacific Ocean Landscape Concept, 73
Panchamrit, 7
Papua New Guinea, 75
Paris, 15
Paris agreement, 9
Passo Fundo, 94
phenomenological analysis, 89
PNG Defense Forces, 72
pollutants, 92
PRC, 68
President Obama, 27
prisoner's dilemma, 177
progressive internationalists, 191
pro-Khalistan, 230

Q

Quad Plus, 72

R

Ravi, Disha, 228
RCP index, 120
reconstitution, 123
Regime Complex for Climate Change, 13, 106

renewables, 140
representative concentration
    pathways, 120
Rio Earth Summit, 47
river ecocide, 83

S
Savar, 94
scaling solar applications for
    agricultural use, 156
scaling solar e-mobility and storage,
    157
second world, 33
Shah, Amit, 239
small island developing states, 68
solar energy, 153
Solar Risk Mitigation Initiative, 156
South Pacific Nuclear Free Zone, 76
space debris, 108
space drag, 108
Stockholm Conference on the Human
    Environment, 171
strategy for sustainable and smart
    mobility, 136
Supply Chain Resilience Initiative,
    78
sustainability, 30
Sustainable Products Policy, 134
synthetic alternatives, 97

T
Taingong space station, 118
Taiwan, 78
Taiwan Straits, 111
tannery industry, 84
third world, 33–34
toolbox, 142
Trans-European Energy Networks
    (TEN-E), 133
Treaty on the Functioning of the
    European Union (TFEU), 132
tropical cyclone, 74

Tropic of Cancer, 163
Tropic of Capricorn, 163
Trump, Donald, 238
tuna resources, 73
Twitter, 222

U
Ukraine war, 141
UN Charter, 33
UNFCCC text of 1992, 27
UN Framework Convention on
    Climate Change (UNFCCC), 131
Union Parishad, 93
United Nations Environment Program
    (UNEP), 228
universalism, 26
urbanization, 6
Ursula von der Leyen, 145
utopian, 142

V
Vietnam War, 84
volcano on Aoba Island, 74

W
walruses, 54
waste dumping, 97
waste management, 97
water contamination, 84
Western documentaries, 61
Westphalian system, 33
whole Earth, 35
World Bank, 31
World Health Organization (WHO), 3
World Resources Institute (WRI), 5
World Solar Technology Summit, 159

X
Xi Jinping, 76

Z
Zero Pollution Action Plan, 137

www.ingramcontent.com/pod-product-compliance
Lightning Source LLC
Chambersburg PA
CBHW050637190326
41458CB00008B/2303